MOLTEN SALT REACTORS AND INTEGRATED MOLTEN SALT REACTORS

Integrated Power Conversion

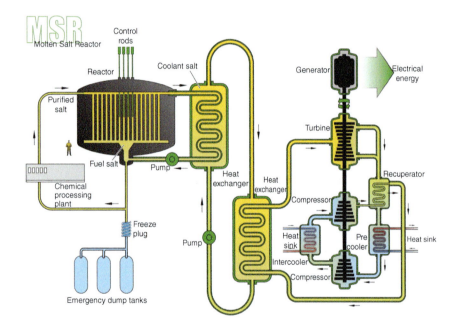

MOLTEN SALT REACTORS AND INTEGRATED MOLTEN SALT REACTORS
Integrated Power Conversion

BAHMAN ZOHURI

Adjunct Professor, Artificial Intelligence Scientist at Golden Gate University, San Francisco, CA; Research Associate Professor, Electrical Engineering and Computer Science, University of New Mexico, Albuquerque, New Mexico, United States of America

ACADEMIC PRESS
An imprint of Elsevier
elsevier.com/books-and-journals

Academic Press is an imprint of Elsevier
125 London Wall, London EC2Y 5AS, United Kingdom
525 B Street, Suite 1650, San Diego, CA 92101, United States
50 Hampshire Street, 5th Floor, Cambridge, MA 02139, United States
The Boulevard, Langford Lane, Kidlington, Oxford OX5 1GB, United Kingdom

Copyright © 2021 Elsevier Inc. All rights reserved.

No part of this publication may be reproduced or transmitted in any form or by any means, electronic or mechanical, including photocopying, recording, or any information storage and retrieval system, without permission in writing from the publisher. Details on how to seek permission, further information about the Publisher's permissions policies and our arrangements with organizations such as the Copyright Clearance Center and the Copyright Licensing Agency, can be found at our website: www.elsevier.com/permissions.

This book and the individual contributions contained in it are protected under copyright by the Publisher (other than as may be noted herein).

Notices
Knowledge and best practice in this field are constantly changing. As new research and experience broaden our understanding, changes in research methods, professional practices, or medical treatment may become necessary.

Practitioners and researchers must always rely on their own experience and knowledge in evaluating and using any information, methods, compounds, or experiments described herein. In using such information or methods they should be mindful of their own safety and the safety of others, including parties for whom they have a professional responsibility.

To the fullest extent of the law, neither the Publisher nor the authors, contributors, or editors, assume any liability for any injury and/or damage to persons or property as a matter of products liability, negligence or otherwise, or from any use or operation of any methods, products, instructions, or ideas contained in the material herein.

British Library Cataloguing-in-Publication Data
A catalogue record for this book is available from the British Library

Library of Congress Cataloging-in-Publication Data
A catalog record for this book is available from the Library of Congress

ISBN: 978-0-323-90638-8

For Information on all Academic Press publications visit our website at
https://www.elsevier.com/books-and-journals

Publisher: Joe Hayton
Acquisition Editor: Katie Hammon
Editorial Project Manager: Madeline Jones
Production Project Manager: Prasanna Kalyanaraman
Cover Designer: Mark Rogers

Typeset by Aptara, New Delhi, India

Dedication

This book is dedicated to my Son Sasha Zohuri, Grandson Dariush Nikpour and my Granddaughter Dony Nikpour, Grandson Kian Emilio Zohuri Caicedo and Firouzeh Arasnia

Contents

About the Author		xiii
Preface		xv
Acknowledgment		xix

1. Molten Salt Reactor History, From Past to Present — 1
   1.1 Introduction 1
   1.2 Aircraft Nuclear Power Reactor Experiment 14
   1.3 Molten Salt Reactor Experiment (MSRE) 19
   1.4 Space-Based Nuclear Reactors 24
   1.5 Sustainable Nuclear Energy 30
   1.6 Prefiltration and Nonprefiltration nuclear reactors 32
   1.7 Nuclear Safeguards 34
   1.8 Safety by Physics Versus by Engineering 37
   1.9 Criticality Issue of Nuclear Energy Systems Driven by MSRs 40
   1.10 Denatured Molten Salt Reactor (DMSR) 43
   1.11 MSR Pros and Cons 46
   1.12 The Potential of the MSR Concept 52
   1.13 Conclusions 53
   References 55

2. Integral Molten Salt Reactor — 59
   2.1 Introduction 59
   2.2 Integral Molten Salt Reactor (IMSR) Descriptions 61
   2.3 Integral Molten Salt Reactor (IMSR) Design 61
   2.4 Integral Molten Salt Safety Philosophy 68
   2.5 Proliferation Defense 72
   2.6 Safety and Security (Physical Protection) 72
   2.7 Description of Turbine-Generator Systems 73
   2.8 Electrical and Integrated and Circuit (I&C) Systems 73
   2.9 Spent Fuel and Waste Management 74
   2.10 Plant Layout 75
   2.11 Plant Performance 75
   2.12 Development Status of Technologies Relevant to the Nuclear Power Plant 76
   2.13 Development Status and Planned Schedule 77
   2.14 Coupling IMSR Technology with Hybrid Nuclear/Renewable Energy Systems 77
      2.14.1 Thermal Storage and Desalination 79
      2.14.2 H_2 from High Temperature Steam Electrolysis 80

	2.14.3	Synthesized Transport Fuels	80
	2.14.4	Ammonia Production Coupled to IMSR	81
	2.14.5	Coupling IMSR Technology into Direct Reduction Steel with H_2	81
2.15	Conclusions		82
References			83

3. New Approach to Energy Conversion Technology — 85

3.1	Introduction	85
3.2	Waste Heat Recovery	91
3.3	PCS Components	91
	3.3.1 Heat Exchangers	91
	3.3.1.1 Recuperative HXs	99
	3.3.1.2 Regenerative HXs	104
	3.3.1.3 Evaporative HXs	105
	3.3.2 Compact HXs	107
3.4	Development of Gas Turbine	110
3.5	Turbomachinery	113
3.6	Heat Transfer Analysis	114
3.7	Combined-Cycle Gas Power Plant	116
3.8	Advanced Computational Materials Proposed for GEN IV Systems	119
3.9	Material Classes Proposed for GEN IV Systems	122
3.10	GEN IV Materials Challenges	122
3.11	GEN IV Materials Fundamental Issues	123
3.12	Capital Cost of Proposed GEN IV Reactors	124
	3.12.1 Economic and Technical of Combined-Cycle Performance	126
	3.12.2 Economic Evaluation Technique	127
	3.12.3 Output Enhancement	129
	3.12.3.1 Gas Turbine Inlet Air Cooling	129
	3.12.3.2 Power Augmentation	130
3.13	Combined-Cycle PCS Driven GEN IV Nuclear Plant	131
	3.13.1 Modeling the Brayton Cycle	132
	3.13.2 Modeling the Rankine Cycle	133
	3.13.3 Results	133
References		135

4. Advanced Power Conversion System Driven by Small Modular Reactors — 137

4.1	Introduction	137
4.2	Currently Proposed Power Conversion Systems for SMRs	141
4.3	Advanced Air-Brayton Power Conversion Systems	142

4.4	Design Equations and Design Parameters		146
	4.4.1	Reactors	146
	4.4.2	Air Compressors and Turbines	147
	4.4.3	Heat Exchanger	150
		4.4.3.1 Primary Heat Exchangers—Sodium-to-Air, Molten Salt-to-Air	151
		4.4.3.2 Economizer—Air to Water	151
		4.4.3.3 Superheaters—Air to Steam	151
		4.4.3.4 Condenser—Steam to Water	152
		4.4.3.5 Recuperator—Air to Air	152
		4.4.3.6 Intercooler—Water to Air	152
	4.4.4	Pumps and Generators	152
	4.4.5	Connections and Uncertainty	152
4.5	Predicted Performance of Small Modular NACC systems		153
4.6	Performance Variation of Small Modular NACC Systems		155
4.7	Predicted Performance for Small Modular NARC Systems		160
4.8	Performance Variation of Small Modular NARC Systems		162
4.9	Predicted Performance for a Small Modular Intercooled NARC System		166
4.10	Performance Variation of Small Modular Intercooled NARC Systems		167
4.11	Conclusions		169
References			169

5. Advanced Nuclear Open Air-Brayton Cycles for Highly Efficient Power Conversion — 171

5.1	Introduction	171
5.2	Background	171
5.3	Combined Cycle Feature	176
5.4	Typical Cycles	176
5.5	Analysis Methodology	177
5.6	Validation of Methodology	180
5.7	Modeling the Nuclear Combined Cycle	181
	5.7.1 Nominal Results for Combined Cycle Model	182
	5.7.2 Extension of Results for Peak Turbine Temperatures	186
5.8	Modeling the Nuclear Recuperated Cycle	186
	5.8.1 Nominal Results for Recuperated Cycle Models	187
	5.8.2 Nominal Results for Recuperated Cycle	190
5.9	Economic Impact	191
5.10	Conclusions	193
References		194

6. Heat pipe driven heat exchangers to avoid salt freezing and control tritium — 197

- 6.1 Introduction — 197
- 6.2 Heat transfer—the traditional application for heat pipes — 198
- 6.3 Prevention of coolant salt freezing — 199
- 6.4 Tritium capture — 199
- 6.5 Reactor systems and heat pipes design requirements — 200
  - 6.5.1 Fluoride-salt-cooled high-temperature reactor — 200
  - 6.5.2 Salt-cooled fusion systems — 202
  - 6.5.3 Molten salt reactors — 203
- 6.6 Salt reactor heat exchanger requirements — 204
- 6.7 Heat pipe design and startup temperature — 209
  - 6.7.1 Choice of fluid — 209
- 6.8 Heat transfer analysis — 213
  - 6.8.1 Heat pipe operation limits — 213
    - 6.8.1.1 Viscous limit — 214
    - 6.8.1.2 Entrainment limit — 214
    - 6.8.1.3 Boiling limit — 215
    - 6.8.1.4 Sonic limit — 216
    - 6.8.1.5 Wicking or capillary limit — 216
  - 6.8.2 Sodium heat pipe experience — 218
- 6.9 Tritium control — 219
- 6.10 Status of technology and path forward — 224
- References — 225

7. Salt cleanup and waste solidification for fission and fusion reactors — 229

- 7.1 Introduction — 229
- 7.2 Requirements — 230
  - 7.2.1 Reactor salt requirements — 231
  - 7.2.2 Molten salt separations requirements — 233
  - 7.2.3 Final waste form requirements — 234
- 7.3 Separations — 236
  - 7.3.1 Distillation — 236
  - 7.3.2 Electrochemical separations — 240
- 7.4 Conversion of salt wastes to high-quality waste forms — 241
  - 7.4.1 Conversion of halide wastes to iron phosphate gas — 241
  - 7.4.2 Conversion of halide wastes to borosilicate glass — 242
- 7.5 Other considerations — 242
- 7.6 Conclusions — 243
- References — 244

Appendix A: A combined cycle power conversion system for small modular LMFBR — 247
- A.1 Introduction — 247
- A.2 The air-Brayton cycle, pros and cons — 247
- A.3 The feed water heater — 248
- A.4 Results — 249
- References — 252

Appendix B: Direct reactor auxiliary cooling system (DRACS) — 253
- B.1 Introduction — 253
- B.2 Decay heat removal system in various reactor designs — 255
- B.3 Experimental validation of passive decay heat removal technology for FHR — 259
- References — 262

Appendix C: Heat pipe general knowledge — 263
- C.1 Heat pipe materials and working fluids — 265
- C.2 Different types of heat pipes — 265
- C.3 Nuclear power conversion — 265
- C.4 Benefits of heat pipe devices — 266
- C.5 Limitations — 266
- C.6 Conclusion — 267
- C.7 Control — 267
- C.8 Engineering — 268
- C.9 Heat pipe applications — 268
  - C.9.1 Heat pipe advantages — 268
  - C.9.2 Heat pipe disadvantages — 269
  - C.9.3 Applications — 269
  - C.9.4 Best applications — 269
  - C.9.5 Possible applications — 269
  - C.9.6 Technology types and resources — 270
  - C.9.7 Efficiency — 270

Appendix D: Variable electricity and steam-cooled based load reactors — 271
- D.1 Introduction — 271
- D.2 Implication of low-carbon grid and renewables on electricity markets — 271
- D.3 Strategies for a zero-carbon electricity grid — 273
- D.4 Nuclear air-Brayton combined cycle strategies for zero-carbon grid — 274
- D.5 Salt-cooled reactors coupled to NACC power system — 274
- D.6 Sodium-cooled reactors (550 °C) coupled to NACC power system — 278
- D.7 Power cycle comparisons — 281
- D.8 Conclusions — 282
- References — 283

Appendix E: Variable electricity and steam-cooled based load reactors   **285**

    E.1 Introduction   285
    E.2 The recuperated Brayton cycle   285
    E.3 Modeling the Rankine cycle   286
    E.4 Computer code running results   286
    E.5 Conclusions   288
    References   288

Index   *289*

About the Author

Dr. Bahman Zohuri currently works for Galaxy Advanced Engineering, Inc., a consulting firm that he started in 1991 when he left both the semiconductor and defense industries after many years working as a chief scientist. After graduating from the University of Illinois in the field of physics, applied mathematics, then he went to the University of New Mexico, where he studied nuclear engineering and mechanical engineering. He joined Westinghouse Electric Corporation, where he performed thermal-hydraulic analysis and studied natural circulation in an inherent shutdown, heat removal system (ISHRS) in the core of a liquid metal fast breeder reactor (LMFBR) as a secondary fully inherent shutdown system for secondary loop heat exchange. All these designs were used in nuclear safety and reliability engineering for a self-actuated shutdown system. He designed a mercury heat pipe and electromagnetic pumps for large pool concepts of an LMFBR for heat rejection purposes for this reactor around 1978 when he received a patent for it. He was subsequently transferred to the defense division of Westinghouse, where he oversaw dynamic analysis and methods of launching and controlling MX missiles from canisters. The results were applied to MX launch seal performance and muzzle blast phenomena analysis (i.e., missile vibration and hydrodynamic shock formation). Dr. Zohuri was also involved in analytical calculations and computations in the study of nonlinear ion waves in rarefying plasma. The results were applied to the propagation of so-called soliton waves and the resulting charge collector traces in the rarefaction characterization of the corona of laser-irradiated target pellets. As part of his graduate research work at Argonne National Laboratory, he performed computations and programming of multiexchange integrals in surface physics and solid-state physics. He earned various patents in areas such as diffusion processes and diffusion furnace design while working as a senior process engineer at various semiconductor companies, such as Intel Corp., Varian Medical Systems, and National Semiconductor Corporation. He later joined Lockheed Martin Missile and Aerospace Corporation as Senior Chief Scientist and oversaw research and development (R&D) and the study of the vulnerability, survivability, and both radiation and laser hardening of different components of the Strategic Defense Initiative, known as Star Wars.

This included payloads (i.e., IR sensor) for the Defense Support Program, the Boost Surveillance and Tracking System, and Space Surveillance and Tracking Satellite against laser and nuclear threats. While at Lockheed Martin, he also performed analyses of laser beam characteristics and nuclear radiation interactions with materials, transient radiation effects in electronics, electromagnetic pulses, system-generated electromagnetic pulses, single-event upset, blast, thermo-mechanical, hardness assurance, maintenance, and device technology.

He spent several years as a consultant at Galaxy Advanced Engineering serving Sandia National Laboratories, where he supported the development of operational hazard assessments for the Air Force Safety Center in collaboration with other researchers and third parties. Ultimately, the results were included in Air Force Instructions issued specifically for directed energy weapons operational safety. He completed the first version of a comprehensive library of detailed laser tools for airborne lasers, advanced tactical lasers, tactical high-energy lasers, and mobile/ tactical high-energy lasers, for example.

He also oversaw SDI computer programs, in connection with Battle Management C3I and artificial intelligence, and autonomous systems. He is the author of several publications and holds several patents, such as for a laser-activated radioactive decay and results of a through-bulkhead initiator. He has published many books and articles with different publishers and publishing company that they all can be found on amazon.com or researchgate.net as well as Internet if the search can be performed under his name.

Preface

The initial goal was an aircraft propulsion reactor, and a molten fluoride-fueled Aircraft Reactor Experiment was operated at Oak Ridge National Laboratory (ORNL) in 1954. In 1956, the objective shifted to civilian nuclear power, and reactor concepts were developed using a circulating UF_4-ThF_4 fuel, graphite moderator, and Hastelloy-N (i.e., corrosion-resistant and high-temperature alloys) pressure boundary. The program culminated in the successful operation of the Molten Salt Reactor (MSR) Experiment from 1965 to 1969. By then the Atomic Energy Commission's goals had shifted to breeder development; the molten-salt program supported on-site reprocessing development and study of various reactor arrangements that had the potential to breed. Some commercial and foreign interest contributed to the program which, however, was terminated by the government in 1976.

ORNL took the lead in researching MSRs through the 1960s. Much of their work culminated with the Molten-Salt Reactor Experiment (MSRE). MSRE was a 7.4 MWth test reactor simulating the neutronic "kernel" of a type of epithermal thorium molten-salt breeder reactor called the Liquid Fluoride Thorium Reactor. The large (expensive) breeding blanket of thorium salt was omitted in favor of neutron measurements.

In 1980, the engineering technology division at ORNL published a paper entitled "Conceptual Design Characteristics of a Denatured Molten-Salt Reactor (DMSR) with Once-Through Fueling." In it, the authors "examine the conceptual feasibility of a molten-salt power reactor fueled with denatured uranium-235 (i.e., with low-enriched uranium) and operated with a minimum of chemical processing." The main priority behind the design characteristics was proliferation resistance. Although the DMSR can theoretically be fueled partially by thorium or plutonium, fueling solely with Low Enriched Uranium helps maximize proliferation resistance.

The MSRE was a prototype for a thorium fuel cycle breeder nuclear power plant. The increased research into Generation IV (GEN-IV) reactor designs renewed interest in the technology. The Generation IV International Forum (GIF) includes "salt processing" as a technology gap for MSRs. The DMSR requires minimal chemical processing because it is a burner rather than a breeder. Both reactors built at ORNL were burner designs. In addition, the choices to use graphite for neutron moderation and enhanced Hastelloy-N for piping simplified the design and reduced R&D. Other important goals of the DMSR were to minimize R&D and to maximize feasibility.

MSR is a class of nuclear fission reactors in which the primary nuclear reactor coolant and/or the fuel is a molten salt mixture. Key characteristics are operation at or close to atmospheric pressure, rather than the 75–150 times the atmospheric pressure of typical light-water reactors (LWRs), hence reducing the large, expensive containment structures used for LWRs and eliminating a source of explosion risk; and higher operating temperatures than in a traditional LWR, hence higher electricity-generation efficiency and in some cases process-heat opportunities. Design challenges include the corrosivity of hot salts and the changing chemical composition of the salt as it is transmuted by reactor radiation.

An MSR was operated at the Critical Experiments Facility of the ORNL in 1957. It was part of the circulating-fuel reactor program of the Pratt & Whitney Aircraft Company. This was called Pratt and Whitney Aircraft Reactor-1 (PWAR-1). The experiment was run for a few weeks and at essentially zero power, although it reached criticality. The operating temperature was held constant at approximately 675°C (1250°F). The PWAR-1 used $NaF\text{-}ZrF_4\text{-}UF_4$ as the primary fuel and coolant. It was one of three critical MSRs ever built.

MSRs use molten fluoride salts as primary coolant, at low pressure. This itself is not a radical departure when the fuel is solid and fixed. But extending the concept to dissolving the fissile and fertile fuel in the salt certainly represents a leap in lateral thinking relative to nearly every reactor operated so far. However, the concept is not new, as outlined below:

1. MSRs operated in the 1960s.
2. They are seen as a promising technology today principally as a thorium fuel cycle prospect or for using spent LWR fuel.
3. A variety of designs is being developed, some as fast neutron types.
4. Global research is currently led by China.
5. Some have solid fuel similar to high-temperature reactor fuel, others have fuel dissolved in the molten salt coolant.

MSRs may operate with epithermal or fast neutron spectrums, and with a variety of fuels. Much of the interest today in reviving the MSR concept relates to using thorium (to breed fissile uranium-233), where an initial source of fissile material such as plutonium-239 needs to be provided. There are a number of different MSR design concepts, and a number of interesting challenges in the commercialization of many, especially with thorium.

The salts concerned as primary coolant, mostly lithium-beryllium fluoride and lithium fluoride, remain liquid without pressurization from about 500°C up to about 1400°C, in marked contrast to a pressurized water reactor, which operates at about 315°C under 150 atmospheres pressure.

The main MSR concept is to have the fuel dissolved in the coolant as fuel salt, and ultimately to reprocess that online. Thorium, uranium, and plutonium all form suitable fluoride salts that readily dissolve in the LiF-BeF$_2$ (FLiBe) mixture, and thorium and uranium can be easily separated from one another in fluoride form. Batch reprocessing is likely in the short term, and fuel life is quoted at 4–7 years, with high burn-up. Intermediate designs and the advanced high-temperature reactor have fuel particles in solid graphite and have less potential for thorium use.

Graphite as a moderator is chemically compatible with fluoride salts.

The Integral Molten Salt Reactor (IMSR) is so-called because it integrates into a compact, sealed, and replaceable nuclear reactor unit called the IMSR core-unit. The core-unit comes in a single size designed to deliver 400 MW of thermal heat which can be used for multiple applications. If used to generate electricity then the notional capacity is 190 MW electrical. The unit includes all the primary components of the nuclear reactor that operate on the liquid molten fluoride salt fuel: moderator, primary heat exchangers, pumps, and shutdown rods.

The core-unit forms the heart of the IMSR system. In the core-unit, the fuel salt is circulated between the graphite core and heat exchangers. The core-unit itself is placed inside a surrounding vessel called the guard vessel. The entire core-unit module can be lifted out for replacement. The guard vessel that surrounds the core-unit acts as a containment vessel. In turn, a shielded silo surrounds the guard vessel.

The IMSR belongs to the DMSR class of MSR. It is designed to have all the safety features associated with the Molten Salt class of reactors including low-pressure operation (the reactor and primary coolant is operated near normal atmospheric pressure), the inability to lose primary coolant (the fuel is the coolant), the inability to suffer a meltdown accident (the fuel operates in an already molten state), and the robust chemical binding of the fission products within the primary coolant salt (reduced pathway for accidental release of fission products).

This book also covers the innovative approach to Nuclear Air Combined Cycle that started by this author and was presented by him in conferences around the world and United States as well as publishing articles around the subject in collaboration with his coauthors Dr. Charles Forsberg of

Massachusetts Institute of Technology and Dr. Patrick J. McDaniel of the University of New Mexico. This approach shows an impactable influence by rising the thermal output efficiency of these MSRs and consequently to bring down the cost of producing electricity from these types of Nuclear Power Plants. Appendices that are written at the end of this book show different techniques and computer modeling that is used to present such an innovative approach.

B. Zohuri
Albuquerque, New Mexico, 2016

Acknowledgment

I am indebted to the many people who aided me, encouraged me, and supported me beyond my expectations. Some are not around to see the results of their encouragement in the production of this book, yet I hope they know of my deepest appreciations. I especially want to thank my true close friends, to whom I am deeply indebted, have continuously given their support without hesitation. They all have always kept me going in the right direction.

Above all, I offer very special thanks to my late mother and father, and to my children, in particular, my son Sasha, daughters, Natasha and Natalie, as well as my grandson Dariush and my granddaughter Donya. They have provided constant interest and encouragement, without which this book would not have been written. Their patience with my many absences from home and long hours in front of the computer to prepare the manuscript are especially appreciated.

CHAPTER 1

Molten Salt Reactor History, From Past to Present

Molten salt reactors (MSRs) are one of six next-generation designs chosen by the Generation IV (GEN IV) program. Traditionally these reactors are thought of as thermal breeder reactors running on the thorium to ^{233}Uranium cycle and the historical competitor to fast breeder reactors. However, simplified versions running as converter reactors without any fuel processing and consuming low-enriched uranium (LEU) are perhaps a more attractive option. Molten-salt reactors were first proposed by Ed Bettis and Ray Briant of Oak Ridge National Laboratory (ORNL) during the post-World War II attempt to design a nuclear-powered aircraft. The attraction of molten fluoride salts for that program was the great stability of the salts, both to high temperatures and to radiation. An active development program aimed at such an aircraft reactor was carried out from about 1950 to 1956.

1.1 Introduction

MSRs are liquid fuel reactors that use solution of uranium and thorium fluorides in lithium and beryllium fluorides as fuels. They operate at high temperature and low pressure and have excellent nuclear characteristics. They offer promise as breeders of fissionable material and as producers of low-cost electricity in large central power stations and recently selected by Department of Energy (DOE) is one of the Six Generation IV power plants and with their high temperature operational level are an excellent candidate as advanced Small Modular Reactor (AdvSMR) utilizing an innovative technology nuclear open air-Brayton combined cycle (NACC) [9].

The origination of an MSR starts with the aircraft reactor experiment (ARE), a small reactor using a circulating molten fuel salt, operated for several days in 1954 and reached a peak temperature of 1620 °F (See Section 1.2 for more details). The aircraft nuclear propulsion (ANP) program and the preceding Nuclear Energy for the Propulsion of Aircraft (NEPA) project worked to develop a nuclear propulsion system for aircraft. The United States Army Air Forces initiated Project NEPA on May 28, 1946 [1]. NEPA operated until May 1951, when the project was transferred to the joint Atomic Energy

Commission (AEC)/USAF ANP [2]. The USAF pursued two different systems for nuclear-powered jet engines, the Direct Air Cycle concept, which was developed by General Electric (GE), and Indirect Air Cycle, which was assigned to Pratt & Whitney. The program was intended to develop and test the Convair X-6 but was cancelled in 1961 before that aircraft was built. The total cost of the program from 1946 to 1961 was about $1 billion [3].

In 1956 interest in the airplane began to fall off, and Alvin Weinberg, an American nuclear physicist who was the administrator at ORNL during and after the Manhattan Project, wished to see whether the molten fluoride fuel technology that had been developed for the aircraft could be adapted to civilian power reactors. Part of his interest stemmed from the fact that all of the other materials and coolants being suggested for reactors had been anticipated by the reactor design group at the Metallurgical Lab oratory in Chicago during World War II. This was new event.

Considering ANP for a civilian program, this type of reactor for this purpose a lot head start information based on historical data was collected during period of its study for military applications for aircraft propulsion purpose, although Division of Reactor Development of the US AEC of the time never showed much enthusiasm for MSR Program for this purpose.

However, with ORNL effort behind the effort of pushing the MSR Program concept, was that when AEC eliminated a few reactor concepts, the decision was to establish task forces of outside experts to evaluate the reactor concepts and, especially, to point out their weaknesses. After a couple of other reactor concepts had been eliminated by this process, the AEC formed the fluid fuels reactors Task Force to evaluate and compare three different fluid fuel reactors: the aqueous homogeneous, the liquid bismuth, and the MSRs. The task force met In Washington for about 2 months early in 1959. In this meeting representative from ORNL were presenting the two reactors concepts one being molten salt system and the aqueous homogeneous, while someone from Brookhaven National Laboratory represented the bismuth-graphite reactor.

Task force members came from other AEC laboratories, from electric utilities, from architect engineering firms, and from the AEC itself. The first sentence of the Summary of the Task Force Report (TID-8505) was, "The Molten Salt Reactor has the highest probability of achieving technical feasibility."1

This conclusion arose from the fact that the molten fluoride salts (1) have a wide range of solubility of uranium and thorium, (2) are stable thermodynamically and do not undergo radiolytic decomposition, (3) have a very low vapor pressure at operating temperatures, and (4) do not attack the nickel-based alloy used in the circulating salt system.

As a result of the task force deliberations, the other two concepts were abandoned, and the molten salt system continued its precarious existence. The reactor considered by the Task Force was a converter reactor, not a breeder, and was described as follows [4].

The reference design MSR is a nickel-molybdenum-chromium-iron alloy-8 (now called Hastelloy-N) vessel containing a graphite assembly 12.25 feet in diameter by 12.25 feet high, through which molten salt flows in vertical channels. The fuel salt is a solution composed of 0.3 mole percent UF_4, 13 mole percent ThF_4, 16 mole percent BeF_2, and 70.7 mole percent ^7LiF. The fuel salt is heated from 1075 °F to 1225 °F in the core and is circulated from the reactor vessel to four primary heat exchangers by four fuel pumps. A barren coolant salt is used as the intermediate heat exchange fluid [5].

In 1972 ORNL proposed a major development program that would culminate in the construction and operation of a demonstration reactor called the molten salt breeder experiment (MSRE). The program was estimated to cost a total of $350 million over a period of 11 years.

However, the MSRE was a very successful experiment, in that it answered many questions and posed but a few new ones. Perhaps the most important result was the conclusion that it was quite a practical reactor. It ran for long periods of time, and when maintenance was required, it was accomplished safely and without excessive delay. Also, it demonstrated the expected flexibility and ease of handling the fuel. As mentioned above, it was the first reactor in the world to operate with ^{233}U as the sole fuel, and the highly radioactive ^{233}U used would have been extremely difficult to handle if it had had to be incorporated into solid fuel elements. In preparation for the run with ^{233}U, the ^{235}U was removed from the carrier salt in 4 days by the fluoride volatility process. This process decontaminated the 218 kg of uranium of gamma radiation by a factor of 4×10^9 so that it could be handled without shielding. As an aside, this equipment used for the MSRE was sufficiently large so that it could satisfactorily handle all of the fuel processing needs for a 1000-MWe molten salt converter reactor.

Moreover, three problems did arise during the construction and operation of the MSRE that required further research and development (R&D) and they were as follows:

1. The first was the Hastelloy-N used for the MSRE was subject to a kind of "radiation hardening," due to accumulation of helium at grain boundaries. Later, it was found that modified alloys that had fine carbide precipitates within the grains would hold the helium and restrain this migration to the grain boundaries. Nevertheless, it is still desirable

to design well-blanketed reactors in which the exposure of the reactor vessel wall to fast neutron radiation is limited.
2. The second problem concerned the tritium produced by neutron reactions with lithium. At high temperatures the radioactive tritium, which is, of course, chemically like hydrogen, penetrates metals quite readily, and unless captured in some way, would appear in the steam generators and reach the atmosphere. After considerable development work, it was found that the intermediate salt coolant, a mixture of sodium fluoride and sodium fluoroborate, would capture the tritium and that it could be removed and isolated in the gas purge system.
3. The third problem came from the discovery of tiny cracks on the inside surface of the Hastelloy-N piping for the MSRE. It was found that these cracks were caused by the fission product tellurium. Later work showed that this tellurium attack could be controlled by keeping the fuel on the reducing side. This is done by adjustment of the chemistry so that about 2% of the uranium is in the form of UF3, as opposed to UF_4. This can be controlled rather easily now that good analytical methods have been developed. If the UF_3 to UF_4 ratio drops too low, it can be raised by the addition of some beryllium metal, which, as it dissolves, will rob some of the fluoride ions from the uranium.

The team working at ONRL as result of their efforts did overcome these problems with reasonable solution around them; however, going from concept to production to produce the electricity from such reactor required serious funding far beyond what was expected in early time for MSR concept.

However, those who would have had to approve such a program were already heavily committed to the liquid metal fast breeder reactor and guiding a very expensive development program that would be spending about $400 million each year by 1975. It was asking too much of human nature to expect them to believe that a much less expensive program could be effective in developing a competing system, and the ORNL proposal was rejected. In January 1973, ORNL was directed to terminate MSR development work. However, for some reasons the program was reinstated a year later, and in 1974 ORNL submitted a more elaborate proposal with suitably inflated costs calling for about $720 million to be spent over an 11-year period. This last proposal was also rejected, and in 1976 ORNL was again ordered to shut down the MSR program "for budgetary reasons" [4].

The decision to cut off the funding for the MSR program was supported by an "evaluation" of the molten salt breeder reactor (MSBR) (1972) prepared internally by the Division of Reactor Development and Technology in response to a request from the Office of Science and Technology.

Although the evaluation report contained no overt recommendations, the Conclusions section, after granting some attractive features of the MSR, emphasized the difficulty of solving a number of problems, including those described above. It was stated that after realistic solutions for these problems had been demonstrated, proceeding toward engineering development would require "reasonable assurances that large-scale government and industrial resources can be made available on a continuing basis in light of other commitments to the commercial nuclear power program and higher priority energy developments." This was, of course, true according to the reference by McPherson write-up [4].

However, this evaluation was prepared before the solutions to the tritium evolution and tellurium-cracking problem were known, including the problem on the materials, the tritium and fuel salt chemical processing. However, the scientist and nuclear engineer who are still pushing either MSR or Integral MSR (IMSR) technologies firmly believe that most of the above problems have been settled by the splendid effort of R&D; performed by ORNL.

However, it seems that the DOE is not yet prepared to make a firm decision toward pushing this technology beyond concept and the prototype, although some private companies around United States with private funding such as Kairos Power is pushing the idea of the pebble-bed fluoride-salt-cooled high-temperature reactor (PB-FHR) that uses FLiBe ($^7Li_2BeF^4$) salt because of its excellent neutronic and thermal-hydraulic properties (See Chapter 6 of this book), which is type of MSR, while Canadian company such as Terrestrial Energy Inc. (TEI) is front runner for concept of integral MSR (IMSR).

Note that the term "MSR" refers to nuclear reactors that use molten salts to transfer heat away from the reactor core. The heat can then be used either to produce electricity or for industrial processes. The use of molten salts to cool the reactor distinguishes MSRs from the other reactor types which use liquid metal, gas, or water as coolants.

MSRs fall into two classes:
- *Salt-cooled reactors*: in which the core contains a solid fuel and liquid salt coolant, and
- *Salt-fueled reactors*: in which the fuel is dissolved within the salt. The term "FHR" was adopted in 2010 to distinguish fluoride salt-cooled MSRs from other MSRs.

Current ORNL concept MSRs use solid fuel, although ORNL has a long history working with salt-fueled MSRs

Furthermore, MSR designs have several inherent safety advantages. The first, and possibly the most important, is that the reactor is operated at low pressure because the coolants never approach boiling point. Even in an accident, there would be no force expelling materials from the reactor, and no high-pressure containment system would be required to prevent such a release.

In addition, under accident conditions MSRs can rely on convection currents—otherwise known as natural circulation—to circulate the cooling salts. This passive safety feature relies on the fact that hot liquids naturally rise and cooler ones sink. Coolant will therefore continue to circulate through the reactor and remove excess heat indefinitely, even if power is lost to the reactor.

While there are no operating MSRs at present, interest in this promising technology can be found both within the United States and internationally. Here are a few examples:

- ORNL has signed a Cooperative Research and Development Agreement (CRADA) with China's Shanghai Institute of Applied Physics (SINAP) to study and develop advanced reactor technologies that use lithium-beryllium-fluoride salts for cooling. This collaboration will initially focus on SINAP's Thorium MSR—Solid Fuel 1 (TMSR-SF1).
- ORNL has developed two FHR-class MSR concepts: the 1500 MW electric advanced high-temperature reactor (AHTR) and the 125 MW thermal small modular AHTR.
- ORNL has entered into an agreement with the Canadian company TEI) to perform a high-level design review of TEI's IMSR.
- In addition, US universities, including the Massachusetts Institute of Technology, the Georgia Institute of Technology, and the University of California–Berkeley, have begun to investigate alternate FHR design concepts programs for universities sponsored by the DOE. On behalf of DOE, ORNL serves to advise and coordinate the research at these universities.

A typical schematic of FHR reactor is illustrated in Fig. 1.1.

In summary, as depicted in Fig. 1.2, An MSR is a class of nuclear fission reactor in which the primary nuclear reactor coolant and/or the fuel is a molten salt mixture.

Key characteristics are operation at or close to atmospheric pressure, rather than the 75–150 times atmospheric pressure of typical light-water reactors (LWR), hence reducing the large, expensive containment

Molten salt reactor history, from past to present 7

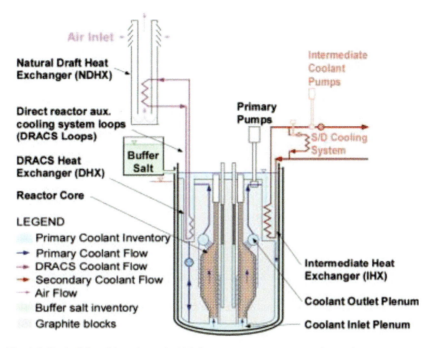

Fig. 1.1 Typical fluoride-salt-cooled high-temperature reactor schematic.

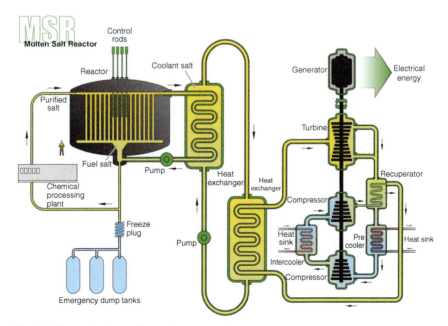

Fig. 1.2 Example of a molten salt reactor scheme.

structures used for LWRs and eliminating a source of explosion risk; and higher operating temperatures than in a traditional LWR, hence higher electricity-generation efficiency and in some cases process-heat opportunities. Design challenges include the corrosivity of hot salts and the changing chemical composition of the salt as it is transmuted by reactor radiation.

While many design variants have been proposed, there are three main categories regarding the role of molten salt, and they are listed in Table 1.1.

The above table has listed all MSR types and each one of them briefly described below according to Wikipedia.com [6] and we have listed them here as well.

(The use of molten salt as fuel and as coolant are independent design choices—the original circulating-fuel-salt MSRE and the more recent static-fuel-salt stable salt reactor (SSR) use salt as fuel and salt as coolant; the dual fluid reactor (DFR) uses salt as fuel but metal as coolant; and the liquid-salt very-high-temperature reactor (FHR) has solid fuel but salt as coolant).

Also, note that, MSR, now often termed liquid fluoride reactors (LFRs), come in many potential forms.

MSRs offer multiple advantages over conventional nuclear power plants (NPPs), although for historical reasons [7] they have not been deployed.

The concept was first established in the 1950s. The early ARE (i.e., See Section 1.2 of this Chapter) was primarily motivated by the small size that the technique offered, while the Molten-Salt Reactor Experiment (MSRE) (i.e., See 1.3 of this Chapter) was a prototype for a thorium fuel cycle breeder NPP. The increased research into GEN-IV reactor designs renewed interest in the technology [8].

In summary, MSR was first investigated as a means of providing a compact, high-temperature power plant for nuclear powered aircraft. As we said at the beginning of this section, in 1954, an ARE (i.e., See Section 1.2 of this chapter) was conducted at ORNL which demonstrated the nuclear

Table 1.1 List of all of types reactors using molten salt fuel.

Category	Examples
Molten salt fuel—circulating	(1) ARE, (2) MSRE, (3) DMSR, (4) MSFR, (5) LFTR, (6) IMSR, (7) AWB, (8) CMSR, (9) EVOL, (10) DFR, (11) TMSR-500
Molten salt fuel—static	(12) SSR
Molten salt coolant only	(13) TMSR (14) FHR

feasibility. Fuel entered the ARE core at 1200 °F and left at 1500 °F when the reactor power level was 2.5 MW.

Now, we can describe every types of reactor using molten salt as fuel and moderator listed in Table 1.1 can be presented as follows:

1. **Aircraft reactor experiment (ARE)**

   MSR research started with the US ARE in support of the US ANP program. ARE was a 2.5 MW_{th} nuclear reactor experiment designed to attain a high energy density for use as an engine in a nuclear-powered bomber.

   The project included experiments, including high temperature and engine tests collectively called the heat transfer reactor experiments: HTRE-1, HTRE-2, and HTRE-3 at the National Reactor Test Station (now Idaho National Laboratory (INL)) as well as an experimental high-temperature MSR at ORNL —The ARE.

   ARE used molten fluoride salt $NaF-ZrF_4-UF_4$ (53–41–6 mol%) as fuel, moderated by Beryllium Oxide (BeO). Liquid sodium was a secondary coolant.

   The experiment had a peak temperature of 860 °C. It produced 100 MWh over 9 days in 1954. This experiment used Inconel 600 alloy for the metal structure and piping.

2. **MSR experiment (MSRE)**

   ORNL took the lead in researching MSRs through the 1960s. Much of their work culminated with the Molten-Salt Reactor Experiment (MSRE). MSRE was an 8 MW_{th} test reactor simulating the neutronic "kernel" of a type of epithermal thorium molten salt breeder reactor called the liquid fluoride thorium reactor (LFTR). The large (expensive) breeding blanket of thorium salt was omitted in favor of neutron measurements.

   MSRE's piping, core vat, and structural components were made from Hastelloy-N, moderated by pyrolytic graphite. It went critical in 1965 and ran for 4 years. Its fuel was $LiF-BeF_2-ZrF_4-UF_4$ (65–29–5–1). The graphite core moderated it. Its secondary coolant was FLiBe ($^2LiF-BeF_2$). It reached temperatures as high as 650 °C and achieved the equivalent of about 1.5 years of full power operation. More details on this reactor are provided in Section 1.3.

3. **ORNL DMSR**

   In 1980, the engineering technology division at ORNL published a paper entitled "Conceptual Design Characteristics of a Denatured Molten-Salt Reactor with Once-Through Fueling." In it, the authors "examine the conceptual feasibility of a molten-salt power reactor fueled with denatured uranium-235 (i.e., with LEU) and operated with a minimum of chemical processing." The main priority behind the design characteristics was proliferation resistance[12]. Although the DMSR can theoretically be fueled partially by thorium or plutonium, fueling solely with LEU helps maximize proliferation resistance.

Other important goals of the DMSR were to minimize R&D and to maximize feasibility. The Gen IV international forum (GIF) includes "salt processing" as a technology gap for MSRs [13]. The DMSR requires minimal chemical processing because it is a burner rather than a breeder. Both reactors built at ORNL were burner designs. In addition, the choices to use graphite for neutron moderation and enhanced Hastelloy-N for piping simplified the design and reduced R&D.

For more details refer to Section 1.8.

4. **Molten salt fast reactor (MSFR)**

The United Kingdom's Atomic Energy Research Establishment (AERE) was developing an alternative MSR design across its National Laboratories at Harwell, Culham, Risley, and Winfrith. AERE opted to focus on a lead-cooled 2.5 GWe MSFR concept using a chloride [14]. They also researched helium gas as a coolant [15, 16].

The UK MSFR would be fueled by plutonium, a fuel considered to be "free" by the program's research scientists, because of the UK's plutonium stockpile.

Despite their different designs, ORNL and AERE maintained contact during this period with information exchange and expert visits. Theoretical work on the concept was conducted between 1964 and 1966, while experimental work was ongoing between 1968 and 1973. The program received annual government funding of around £100,000–£200,000 (equivalent to £2m–£3m in 2005). This funding came to the end in 1974, partly due to the success of the Prototype Fast Reactor at Dounreay, which was considered a priority for funding as it went critical in the same year [14].

5. **Liquid fluoride thorium reactor (LFTR)**

INL in United Sates designed a molten salt–cooled, molten salt–fueled reactor with a prospective output of 1000 MW_e [17].

Kirk Sorensen, former NASA scientist and chief nuclear technologist at Teledyne Brown Engineering, is a long-time promoter of the thorium fuel cycle, coining the term LFTR. In 2011, Sorensen founded Flibe Energy, a company aimed at developing 20–50 MW LFTR reactor designs to power military bases. (It is easier to approve novel military designs than civilian power station designs in the US nuclear regulatory environment) [18-21].

Transatomic Power pursued what it termed a waste-annihilating MSR (acronym WAMSR), intended to consume existing spent nuclear fuel [22], from 2011 until ceasing operation in 2018 [23].

In January 2016, the United States DoE announced an $80m award fund to develop Gen IV reactor designs [24]. One of the two beneficiaries, Southern Company will use the funding to develop a molten chloride fast reactor, a type of MSR developed earlier by British scientists [14].

6. Integral molten salt reactor (IMSR)

Terrestrial Energy, a Canadian-based company, is developing a DMSR design called the IMSR (i.e., See Chapter 2 of this book for more information). The IMSR is designed to be deployable as a small modular reactor (SMR). Their design currently undergoing licensing is 400 MW thermal (190 MW electrical). With high operating temperatures, the IMSR has applications in industrial heat markets as well as traditional power markets. The main design features include neutron moderation from graphite, fueling with LEU, and a compact and replaceable Core-unit. Decay heat is removed passively using nitrogen (with air as an emergency alternative). The latter feature permits the operational simplicity necessary for industrial deployment [25].

Terrestrial completed the first phase of a prelicensing review by the Canadian Nuclear Safety Commission (CNSC) in 2017, which provided a regulatory opinion that the design features are generally safe enough to eventually obtain a license to construct the reactor [26].

7. Atomic waste burner (AWB)

In Denmark, the Copenhagen Atomic is a Danish molten salt technology company developing mass manufacturable MSRs. The Copenhagen AWB is a single-fluid, heavy water moderated, fluoride-based, thermal spectrum. and autonomously controlled MSR. This is designed to fit inside of a leak-tight, 40-foot, stainless steel shipping container. The heavy water moderator is thermally insulated from the salt and continuously drained and cooled to below 50 °C. A molten lithium-7 deuteroxide (^{7}LiOD) moderator version is also being researched. The reactor utilizes the thorium fuel cycle using separated plutonium from spent nuclear fuel as the initial fissile load for the first generation of reactors, eventually transitioning to a thorium breeder [27]. Copenhagen Atomics is actively developing and testing valves, pumps, heat exchangers, measurement systems, salt chemistry and purification systems, and control systems and software for molten salt applications [28].

8. **Compact molten salt reactor (CMSR)**
   In Denmark Seaborg Technologies is developing the core for a Compact MSR (CMSR). The CMSR is a high-temperature, single-salt, thermal MSR designed to go critical on commercially available LEU. The CMSR design is modular and uses proprietary NaOH moderator [29]. The reactor core is estimated to be replaced every 12 years. During operation, the fuel will not be replaced and will burn for the entire 12-year reactor lifetime. The first version of the Seaborg core is planned to produce 250 MW_{th} power and 100 MW_e power. As a power plant, the CMSR will be able to deliver electricity, clean water, and heating/cooling to around 200,000 households [30].

9. **Evaluation and viability of liquid (EVOL) fuel fast reactor system**
   French Centre national de la recherche scientifique (CNRS) or National Center for Scientific Research is the French state research organization and is the largest fundamental science agency in Europe. The CNRS project EVOL (EVOL fuel fast reactor system) project, with the objective of proposing a design of the MSFR [31], released its final report in 2014 [32]. Various MSR projects such as FHR, MOSART, MSFR, and TMSR have common R&D themes [33].
   The EVOL project will be continued by the EU-funded Safety Assessment of the MSFR (SAMOFAR) project, in which several European research institutes and universities collaborate [34].

10. **Dual fluid reactor (DFR)**
    The German Institute for Solid State Nuclear Physics in Berlin has proposed the DFR as a concept for a fast breeder, lead-cooled MSR. The original MSR concept used the fluid salt to provide the fission materials and also to remove the heat. Thus, it had problems with the needed flow speed. Using two different fluids in separate circles solves the problem.

11. **Thorium molten salt reactor (TMSR-500)**
    The Thorium MSR (TMSR-500) is an NPP being designed for the Indonesian market by ThorCon. The TMSR-500 is based on an SMR that employs molten salt technology. The reactor design is based on the DMSR design from ORNL and employs liquid fuel, rather than a conventional solid fuel. The liquid contains the nuclear fuel and serves as primary coolant [35]. ThorCon plans to manufacture the complete

power plants cheaply in shipyards employing modern shipbuilding construction techniques.

ThorCon is developing the TMSR-500 MSR for the Indonesian market.

12. **Stable salt reactor (SSR)**

    The SSR is a relatively recent concept that holds the molten salt fuel statically in traditional LWR fuel pins. Pumping of the fuel salt, and all the corrosion/deposition/maintenance/containment issues arising from circulating a highly radioactive, hot, and chemically complex fluid, are no longer required. The fuel pins are immersed in a separate, nonfissionable fluoride salt that acts as a primary coolant.

13. **Thorium molten salt reactor (TMSR)**

    China initiated a TMSR research project in January 2011 [36]. A 100 MW demonstrator of the TMSR-solid fuel version (TMSR-SF), based on pebble-bed technology, planned to be ready by 2024. Initially, a 10 MW pilot and a larger demonstrator of the TMSR-liquid fuel (TMSR-LF) variant were targeted for 2024 and 2035, respectively [37, 38]. China then accelerated its program to build two 12 MW reactors underground at Wuwei research facilities in Gansu Province by 2020 [39]. Heat from the thorium molten-salt reaction would be used to produce electricity, hydrogen, industrial chemicals, desalination, and minerals [39]. The project also seeks to test new corrosion-resistant materials [39].

    In 2017, ANSTO/Shanghai Institute Of Applied Physics announced the creation of a NiMo-SiC alloy for use in MSRs [40, 41].

14. **Fluoride salt-cooled high-temperature reactor (FHR)**

    The FHR, also known as liquid-salt Very High-Temperature Reactor (LS-VHTR).

    This approach involves using a fluoride salt as the coolant. Both the traditional MSR and the very-high-temperature reactor (VHTR) were selected as potential designs for study under the Gen IV Initiative (GEN-IV). One version of the VHTR under study was the LS-VHTR, also commonly called the AHTR.

It uses liquid salt as a coolant in the primary loop, rather than a single helium loop. It relies on "tri-structural isotropic (TRISO)" fuel dispersed in graphite. Early AHTR research focused on graphite in the form of graphite rods that would be inserted in hexagonal moderating graphite blocks, but current studies focus primarily on pebble-type fuel. The LS-VHTR can work at very high temperatures (the boiling point of most molten salt candidates is >1400 °C); low-pressure cooling that can be used to match hydrogen production facility conditions (most thermochemical cycles require temperatures in excess of 750 °C); better electric conversion efficiency than a helium-cooled VHTR operating in similar conditions; passive safety systems and better retention of fission products in the event of an accident

1.2 Aircraft Nuclear Power Reactor Experiment

The concept of nuclear-powered aircraft was first formally studied in May 1946 by the US Army Air Forces. It was supposed that the unique characteristics of nuclear power could be applied to long-range supersonic flight, which was considered highly valuable in terms of military strategy. Challenges in the proposal were understood immediately, and by 1950 the Atomic Energy Commission joined with the Air Force to study the possibilities via technology development in the ANP program.

The ORNL staff of the ANP project decided that technical information and experience needed to support the objective of nuclear-powered flight could most economically be obtained from building and operating the ARE. They considered the task of flying a supersonic airplane on nuclear energy exceedingly complex and thought more than one experimental reactor may be necessary before sufficient information was obtained to design and construct a reactor for flight.

Originally, the ARE was conceived as a liquid sodium metal–cooled BeO-moderated solid-fuel reactor. The BeO moderator blocks were purchased with the solid-fuel design in mind. However, concerns regarding chain reaction stability related to xenon in solid fuel at very high temperatures were serious enough to warrant abandoning solid fuel and replacing it with circulating fluid fuel. A fluid-fueled option with molten fluoride salt was worked into the original design.

The ARE as its cross-section illustrated in Fig. 1.3, was an experimental nuclear reactor designed to test the feasibility of fluid-fuel, high-temperature, high-power-density reactors for the propulsion of supersonic aircraft. It operated between November 8 and 12, 1954 at the ORNL with a maximum sustained power of 2.5 MW and generated a total of 96 MW-h of energy.

The ARE was the first reactor to use a circulating molten salt fuel. The hundreds of engineers and scientists working on ARE provided technical data, facilities, equipment, and experience that enabled the broader development of MSR as well as liquid metal–cooled reactors.

The heat exchanger chamber took up significantly more space than the reactor and dump tank chambers. Development of pumps, heat exchangers,

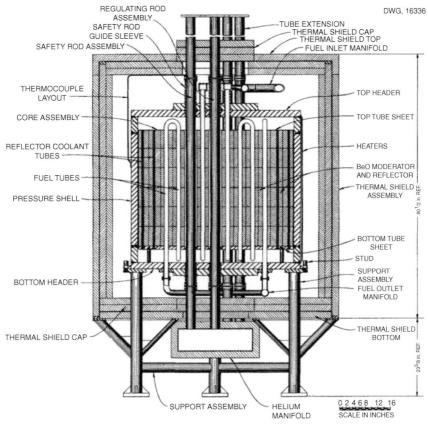

Fig. 1.3 A cross-section of the aircraft reactor experiment (ARE). *(Courtesy of Oak Ridge National Laboratory).*

valves, pressure instrumentation, and cold traps spanned from late 1951 to summer 1954. Much of the work was based on extensive experience at lower temperature from Argonne National Laboratory (ANL) and the Knolls Atomic Power Laboratory.

The ambitious goals and military importance of the ANP catalyzed a significant amount of research and development of complex systems in challenging high-temperature, high-radiation environments.

Corrosion and hot sodium handing studies began in 1950. Investigations of the engineering and fabrication problems involved in handling molten fluoride salts began in 1951 and continued through 1954. Natural-convection corrosion test loops (Fig. 1.4) were operated to down-select suitable material and fuel combinations. Subsequent studies in forced-circulation test loops established means to minimize corrosion and mass transfer.

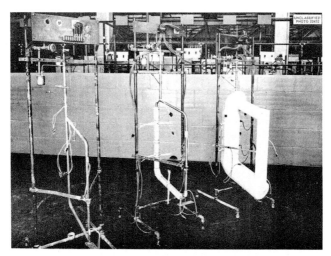

Fig. 1.4 Natural-convection corrosion test loops. *(Courtesy of Oak Ridge National Laboratory)*.

Techniques had to be developed concerning the construction, preheating, instrumentation, and insulation of reliable leak-tight high-temperature circuits made of Inconel. They found that all-welded construction was necessary.

In all, equipment development in support of high-temperature leak-tight operation lasted about 4 years.

Natural-convection corrosion test loops used during the ANP project to down-select workable fuel/material combinations.

The ARE was operated successfully. It became critical with a mass of 32.8 lbs (14.9 kg) Uranium-235. It was very stable as a result of its strong negative fuel temperature coefficient (measured at -9.8×10^{-5} dk/k/°F. (Fig. 1.5).

The assembly was first sufficiently assembled on August 1, 1954, at which point a three-shift operation commenced for tests. Hot sodium metal was flowed through the system beginning on September 26 to test the process equipment and instrumentation. Problems with the sodium vent and sodium purification systems required lengthy repairs.

After several sodium dumps and recharges, carrier salt was introduced to the system on October 25. Fuel was first added to the reactor on October 30, 1954. Initial criticality was reached at 3:45 PM on November 3, after a painstaking and careful process of adding the enriched fuel. Much of the 4 days was spent removing plugs and repairing leaks in the enrichment line.

Fig. 1.5 Aircraft reactor experiment full-scale mockup with clear tubes. *(Courtesy of Oak Ridge National Laboratory).*

The ARE successfully demonstrated the feasibility of generating heat by fission in a fused-fluoride circulating fuel. Most of the heat was removed from the reactor by the fused fluoride at 1580 °F. Sodium at 1350 °F was used to cool the BeO moderator. With minor exceptions all the components proved to be adequate.

As we stated, the ARE required more than 4 years of research and development and is believed to be the first reactor to generate nuclear power above 1500 °F.

The development of components and fabrication techniques for this reactor consumed a 4-year period, during which time the technology for handling high-temperature fluids was extended to equipment operable above 1500 °F. The methods used for determining compatibility of materials under static and dynamic conditions, standards for materials, techniques for welding, fabrication, and assembly and the design criteria for pumps, seals, valves, heat exchangers, cold traps, expansion tanks, instrumentation, preheating devices, insulation, etc. like molten salt fuel pump that was used, are illustrated in Fig. 1.6.

It is presumed that the reader is familiar with the basic concepts of the design, the physics, the chemistry, and the metallurgy which led to this particular reactor system, as these topics have been covered in other reports [58] as has the operation of the reactor experiment.

As part of ANP program, the ANP and the preceding NEPA project worked to develop a nuclear propulsion system for aircraft. The United

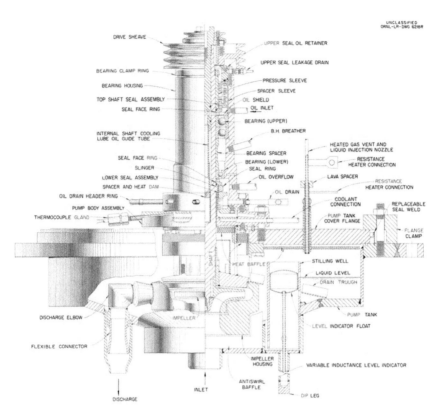

Fig. 1.6 The molten salt fuel pump used in the aircraft reactor experiment. *(Courtesy of Oak Ridge National Laboratory).*

States Army Air Forces initiated Project NEPA on May 28, 1946 [59]. NEPA operated until May 1951, when the project was transferred to the joint AEC/USAF ANP [59]. The USAF pursued two different systems for nuclear-powered jet engines, the Direct Air Cycle concept, which was developed by General Electric, and Indirect Air Cycle, which was assigned to Pratt & Whitney. The program was intended to develop and test the Convair X-6 but was cancelled in 1961 before that aircraft was built.

The total cost of the program from 1946 to 1961 was about $1 billion.

Decommissioning of this program and the full system was carefully dismantled in February 1955.

The main fuel pump bowl surveyed at 900 mrem/h at 5 ft. A portable grinder that could be operated from within a lead box was built to cut the fuel lines near the reactor can. Once it was free, the reactor was moved to storage and later to a burial ground. The fuel in the dump tank was slated to be reprocessed.

About 60 samples of equipment and material were taken for detailed analysis and examination. Metallographic, activation, visual, stereo-photographic, and leak tests were performed.

1.3 Molten Salt Reactor Experiment (MSRE)

The Molten-Salt Reactor Experiment (MSRE) was an experimental MSR at the ORNL researching this technology through the 1960s; constructed by 1964, it went critical in 1965 and was operated until 1969 [10].

The MSRE is an 8-MW(thermal) reactor in which molten fluoride salt at 1200 °F circulates through a core of graphite bars. Its purpose was to demonstrate the criticality of the hey features of molten-salt power reactors.

Operation with ^{235}U (33% enrichment) in the fuel salt began in June 1965, and, by March 1968 nuclear operation amounted to 9000 equivalent full-power hours. The goal of demonstrating reliability had been attained over the last 15 months of ^{235}U operation the reactor had been critical 80% of the time. At the end of a 6-month run which climaxed this demonstration, the reactor was shut down and the 0.9 mole% uranium in the fuel was stripped very efficiently in an on-site fluorination facility. Uranium-233 was then added to the carrier salt, making the MSRE the world's first reactor to be fueled with this fissile material. Nuclear operation was resumed in October 1968, and, over 2500 equivalent full-power hours have now been produced with ^{233}U.

The MSRE has shown that salt handling in an operating reactor is quite practical, the salt chemistry reactor is quite practical, the salt chemistry is well behaved, there is practically no corrosion, the nuclear characteristics are very close to predictions, and the system is dynamically stable. Containment of fission products has been excellent, and maintenance of radioactive components has been accomplished without unreasonable delay and with very little radiation exposure.

The achievement that should strengthen confidence in the practicality of the MSR concept. [10, 11].

The flowsheet of MSRE is presented in Fig. 1.7, which shows the normal operating condition at 8 MW (thermal), the maximum heat removal capability of the air-cooled secondary heat exchanger-type shell in tube. However, the author suggests implementation of a passive heat exchanger utilizing heat pipe heat exchanger. See Chapter 6 of this book.

The physical arrangement of the salt systems is shown in Fig. 1.8. The building housing the reactor is the one in which the ARE was operated in

20 Molten salt reactors and integrated molten salt reactors

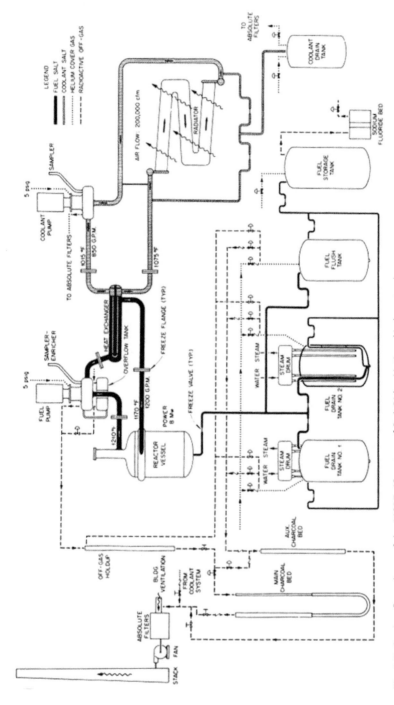

Fig. 1.7 Design flow sheet of the MSRE. *(Courtesy of Oak Ridge National Laboratory)*.

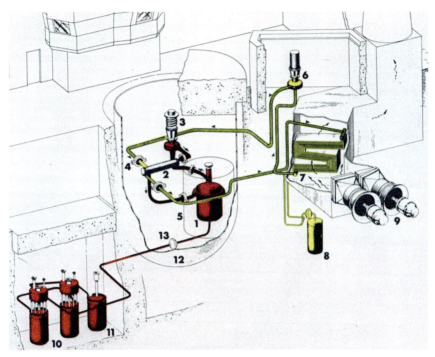

Fig. 1.8 Diagram of molten salt reactor experiment. *(Courtesy of Oak Ridge National Laboratory).*

1954. The cylindrical reactor cell was added for the Aircraft Reactor Test (which was never built) and was adapted for MSRE use.

The list of each component in the above Fig. 1.8 is as follows:

MSRE plant diagram:
1. Reactor vessel
2. Heat exchanger
3. Fuel pump
4. Freeze flange
5. Thermal shield
6. Coolant pump
7. Radiator
8. Coolant drain tank
9. Fans
10. Fuel drain tanks
11. Flush tank
12. Containment vessel
13. Freeze valve. Also note Control area in upper left and Chimney upper right

Details of the MSRE core and reactor vessel are shown in Fig. 1.9. The 54-inch-diameter core is made up of graphite bars, 2 inch. square and 64 inch. tall, exposed directly to fuel that flows in passages machined into the faces of the bars. The graphite was especially produced to have low permeability, and, as salt does not wet the graphite, very high pressure would be required to force any significant amount of salt into the graphite. Some cracks developed in the manufacture of the graphite, but cracked bars were accepted when tests showed effects attending heating and salt intrusion into cracks were inconsequential.

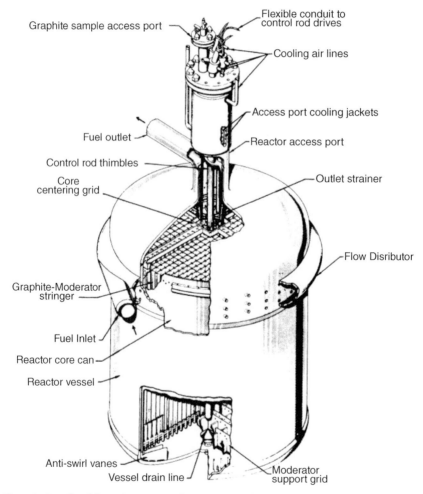

Fig. 1.9 Details of the MSRE core and reactor vessel.

All metal components in contact with molten salt are made of Hastelloy-N (formerly called INOR-8). Metal corrosion by salt mixtures consists of oxidation of metal constituents to their fluoride salts, which do not form protective films. Attack is therefore limited only by the thermodynamic potential for the oxidation reaction, and is selective, removing the least-noble constituent, which in the case of Hastelloy-N is chromium.

However, the diffusion coefficient of the chromium in the metal is such that there is practically no chromium leaching at temperatures below 1500 °F. Impurities in the salt, such as FeF_2, react with Hastelloy-N, but this is a limited effect which goes to completion soon after the salts are loaded. The metallurgy and technology of Hastelloy-N have been thoroughly developed and it has been approved for construction under american society of mechanical engineers (ASME) Unfired Pressure Vessel and Nuclear Vessel Codes.

Hastelloy-N is stronger than austenitic stainless steel and most nickel-base alloys but, like these metals it is subject to deterioration of high-temperature ductility and stress-rupture life by neutron irradiation.

These effects are due to accumulation in grain boundaries of helium produced by n, α reactions. In the MSRE neutron spectrum the fast neutron reactions with nickel are insignificant compared to the slow neutron reactions with impurity boron. Careful analysis of stresses and neutron fluxes in the molten salt reactor experiment unit (MSREU) led to the conclusion that the service life of the reactor vessel would extend at least 20 000 h beyond the point at which the properties of the metal began to be seriously affected by the neutron exposure [10, 11].

The control rods are flexible, consisting of hollow cylinders of Gd_2O_3-Al_2O_3 ceramic, canned in Inconel and threaded on a stainless-steel hose which also serves as a cooling-air conduit. An endless-chain mechanism, driven through a clutch, raises and lowers the rods at 0.5 inch./s.

When scrammed, the rods fall with an acceleration of -12 ft/s². The bowl of the fuel pump is the surge space for the circulating loop. Dry, deoxygenated helium at 50 psig blankets the salt in the pump bowl.

About 50 of the 120 gal/min discharged by the pump is sprayed into the gas space to provide contact between salt and cover gas to allow ^{135}Xe to escape from the salt. (The solubility of xenon and krypton in the salt is very low). A flow of 4 L/min STP of helium carries the xenon and krypton out of the pump bowl, through a holdup volume providing ~40-min delay, a filter station, and a pressure-control valve to charcoal beds.

The charcoal beds consist of pipes filled with charcoal, submerged in a water-filled pit at ~90 °F. The beds are operated on a continuous flow basis and delay xenon for ~90 days and krypton for ~7 days. Thus, only stable or long-lived gaseous nuclides are present in the helium, which is discharged through a stack after passing through the beds.

All salt piping and vessels are electrically heated to prepare for salt filing and to keep the salt molten when there is no nuclear power. In the reactor and drain-tank cells, where radiation levels make remote maintenance necessary, heater elements and reflective metal insulation are combined in removable units. Thermocouples under each heater monitor temperatures to avoid overheating the empty pipe. The radiator is equipped with doors that drop to block the air duct and seal the radiator enclosure if the coolant salt circulation stops and there is danger of freezing salt in the tubes.

More details can be found in references of 10 and 11.

1.4 Space-Based Nuclear Reactors

Spaced-based nuclear reactors are seriously are considered as remote power sources. Among countries with capability of reactor building technologies such as China and Russia have investigated space-based nuclear systems for more than a half-century and the United States has done so intermittently [44, 45].

In the United Sates, the Westinghouse eVinci reactors would be fully factory built and fueled. As well as power generation, process heat to 600 °C would be available. Units would have 5- to 10-year operational lifetime, with walk-away safety due to inherent feedback diminishing the nuclear reaction with excess heat, also effecting load-following. However, eVinci is not considered as an AdvSMR, due to its small scale size it is a good candidate for space-based source of energy as mentioned above and falls into category of micro-reactor for a lack of better classification choice. As it is illustrated in Fig. 1.10, the eVinci micro reactor's innovative design is a combination of nuclear fission, space reactor technologies, and 50+ years of commercial nuclear systems design, engineering, and innovation. The eVinci microreactor aims to create affordable and sustainable power with improved reliability and minimal maintenance, particularly for energy consumers in remote locations. The small size of the generator allows for easier transportation and rapid, on-site installation in contrast to large, centralized stations. The reactor core is designed to run for more than 10 years, eliminating the need for frequent refueling.

The key benefits of the eVinci microreactor are attributed to its solid core and advanced heat pipes. The core encapsulates fuel to significantly

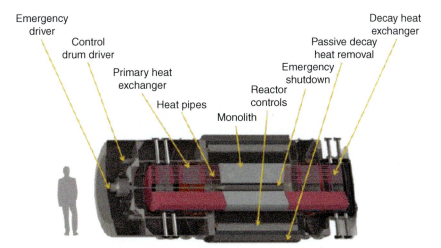

Fig. 1.10 Westinghouse eVinci heat pipe reactor.

reduce proliferation risk and enhances overall safety for the user. The heat pipes enable passive core heat extraction and inherent power regulation, allowing autonomous operation and inherent load following capabilities. These advanced technologies together make the eVinci microreactor a pseudo "solid-state" reactor with minimal moving parts.

The key attributes of eVinci microreactor are as follows:
- Transportable energy generator.
- Fully factory built, fueled and assembled.
- Combined heat and power—200–25 MWe.
- Up to 600 °C process heat.
- 5- to 10-year life with walkaway inherent safety.
- Target less than 30 days onsite installation.
- Autonomous load management capability.
- Unparalleled proliferation resistance.
- High reliability and minimal moving parts.
- Green field decommissioning and remediation.

As part of feature of eVinci as a microreactor design it could represent itself as an ultimate energy solution for the off-grid customer as depicted in Fig. 1.11.

Use of the heat pipes in a reactor system addresses some of the most difficult reactor safety issues and reliability concerns present in current Generations II and III and to some extent, Gen IV concepts commercial nuclear reactors—in particular, loss of a primary coolant. Heat pipes operate in a passive mode at relatively low pressures, less than an atmosphere.

Fig. 1.11 Various application of eVinci in off-grid mode.

Each individual heat pipe contains only a small amount of working fluid, which is fully encapsulated in a sealed steel pipe. There is no primary cooling loop, hence no mechanical pumps, valves, or large-diameter primary loop piping typically found in all commercial reactors today. Heat pipes simply transport heat from the in-core evaporator section to the ex-core condenser in continuous isothermal vapor/liquid internal flow. Heat pipes offer a new and unique means to remove heat from a reactor core.

The Soviet Union's Radar Ocean Reconnaissance Satellite (RORSAT) (Fig. 1.12) and COMOS (Fig. 1.13) lines Satellites employed nuclear reactors from 1970 to 1988, especially in the BUK mixed oxide mixed oxide fuel (MOX) reactor model.

Note that the Cosmos is a long and ongoing series of Earth satellites launched by Russia for a variety of military and scientific purpose. The first one was placed in orbit on March 16, 1962.

Note that the MOX fuel, commonly referred to as MOX fuel, is nuclear fuel that contains more than one oxide of fissile material, usually consisting of plutonium blended with natural uranium, reprocessed uranium, or depleted uranium. MOX fuel is an alternative to the LEU fuel used in the LWRs that predominate nuclear power generation and it (i.e., LEU) is very appropriate for export licensing of NPP, which satisfying nuclear safeguarding propose as well as proliferation and nonproliferation is concerned.

An important and fundamental aspect of nuclear power is that, instead of just using the prepared nuclear fuel once and then dumping it as waste,

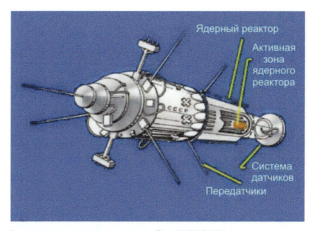

Fig. 1.12 Radar ocean reconnaissance satellite (RORSAT).

Fig. 1.13 An Early COSMOS satellite artistic image.

most if it can be recycled, thus closing the fuel cycle. The current means of doing this is by separating the plutonium and recycling that, mixed with depleted uranium, as MOX fuel. Very little recovered uranium is recycled at present. Another way of closing the fuel cycle is to recycle all the uranium and plutonium without separating them and topping up with some fresh uranium enriched to a higher level than usual. This is regenerated mixture fuel, under development. In each case, the fission products and minor actinides are separated as high-level waste when the used fuel is processed.

See Figs. 1.14–1.16 for additional historical and technical detail for international deployment of space-based nuclear reactors.

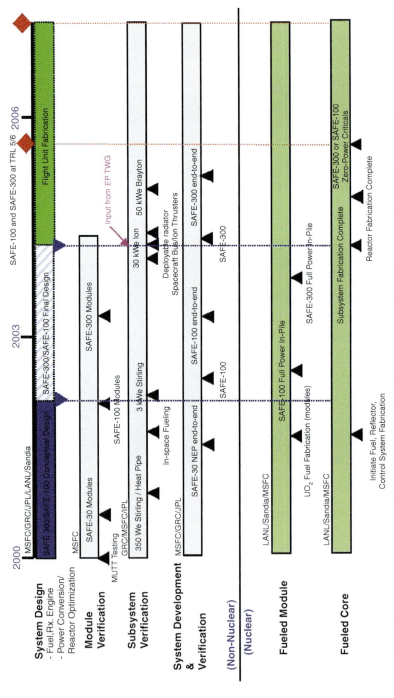

Fig. 1.14 SAFE construction and deployment proposed timeline. *(Source: Adapted from Van Dyke et al. (2001))* [46].

Molten salt reactor history, from past to present   29

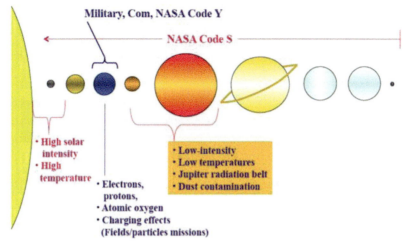

Fig. 1.15 NASA logistical considerations for short- and long-range energy sources (Van Dyke et al., 2001). *(Source: Image courtesy JPL).*

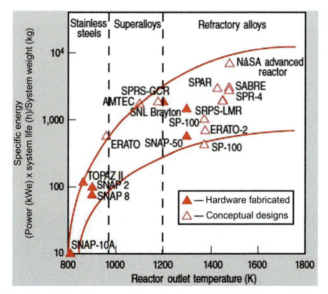

Fig. 1.16 Efforts and concepts, accompanying reactor outlet temperatures and specific energies. *(Courtesy of Elsevier [49]).*

NASA/JPL (jet propulsion laboratory) work has continued on the SAFE nuclear reactor [46–48]) (Fig. 1.10). NASA/JPL have set forth stringent conditions regarding extraterrestrial, lunar, and/or Martian use of nuclear reactors as reactor protection systems (RPSs):

Fig. 1.12 is the efforts and concepts, accompanying production temperatures and specific energies (estimated where appropriate) throughout the decades as categorized by material. Adapted from Yvon and Carre [49]

Nuclear reactors as RPSs can accommodate every need for remote, human exploration.

Note that the RPS is a set of nuclear safety and security components in an NPP designed to safely shut down the reactor and prevent the release of radioactive materials. The system can "trip" automatically (initiating a scram), or it can be tripped by the operators.

1.5 Sustainable Nuclear Energy

Combining conventional and renewable energy systems are the keys to meeting the world's needs for sustainable, reliable, and affordable power. Combating climate change and addressing increasing energy demands are top priorities.

Considering that MSR and its size and footprint infrastructure is designed as SMR by even having a design that is an approach to "*Ultra Compact*" property of NPP is a fantastic source of generating clean energy to produce elasticity at a cheaper cost. (Fig. 1.17).

Some companies such as Seabog of Denmark with their CMSR (i.e., Fig. 1.17) and its proprietary of moderator avoid the use of graphite. Additionally, it gives them three unique advantages listed below.
1. 12 years of operation without refueling
2. Economical at low power (100 MW).
3. Unprecedented compactness, (container-sized reactor module)

Fig. 1.17 Different source of sustainable energy sources. *(Courtesy of Siemens Energy)*.

As we know and have stated throughout this book, there is increasing interest in SMRs and their applications. SMRs are newer generation reactors designed to generate electric power up to 300 MW, whose components and systems can be shop fabricated and then transported as modules to the sites for installation as demand arises. Most of the SMR designs adopt advanced or even inherent safety features and are deployable either as a single or multimodule plant. SMRs are under development for all principal reactor lines: water-cooled reactors, high-temperature gas-cooled reactors, liquid-metal, sodium and gas-cooled reactors with fast neutron spectrum, and MSRs. The key driving forces of SMR development are fulfilling the need for flexible power generation for a wider range of users and applications, replacing ageing fossil-fired units, enhancing safety performance, and offering better economic affordability.

Many SMRs are envisioned for niche electricity or energy markets where large reactors would not be viable. SMRs could fulfill the need of flexible power generation for a wider range of users and applications, including replacing aging fossil power plants, providing cogeneration for developing countries with small electricity grids, remote and off grid areas, and enabling hybrid nuclear/renewables energy systems. Through modularization technology, SMRs target the economics of serial production with shorter construction time. Near-term deployable SMRs will have safety performance comparable or better to that of evolutionary reactor designs.

Though significant advancements have been made in various SMR technologies in recent years, some technical issues still attract considerable attention in the industry. These include for example, control room staffing and human factor engineering for multimodule SMR plants, defining the source term for multimodule SMR plants with regard to determining the emergency planning zone, developing new codes and standards, and also load-following operability aspects. Some potential advantages of SMRs such as the elimination of public evacuation during an accident or a single operator for multiple modules are being challenged by regulators. Furthermore, although SMRs have lower upfront capital cost per unit, their generating cost of electricity will probably be substantially higher than that for large reactors.

Moreover, due to population growth at an annual rate of 17%–19% worldwide, the demand for electricity production is on rise, thus quest for the source of decarbonization of energy production–driven electricity is the main goal of sustainability of such energy both renewable and nonrenewable is in horizon going forward. Nuclear energy can be considered a

32 Molten salt reactors and integrated molten salt reactors

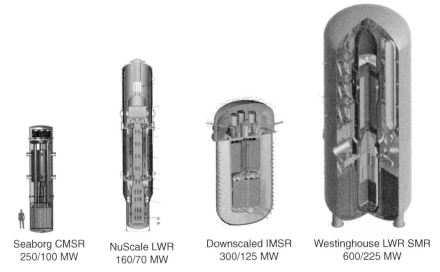

Seaborg CMSR
250/100 MW

NuScale LWR
160/70 MW

Downscaled IMSR
300/125 MW

Westinghouse LWR SMR
600/225 MW

Fig. 1.18 Presentation of different molten salt reactor size designed by different manufactures around the world.

nonrenewable source of energy, thus GEN-IV NPP technology in form of SMR has gain tremendous momentum among nuclear industries. And as we know MSR-type reactor is one of the choices of SMR technology that is getting pushed from concept to production stage (Fig. 1.18).

As it is presented in Fig. 1.17, combining conventional and renewable energy systems are the keys to meeting the world's needs for sustainable, reliable, and affordable power. Combating climate change and addressing increasing energy demands are top priorities. Consequently, electricity for future generation is an element of an affordable and sustainable energy, which is a critical driver for growth and prosperity of a nation.

1.6 Prefiltration and Nonprefiltration nuclear reactors

As part of the World War II effort to develop the atomic bomb, reprocessing technology was developed to chemically separate and recover fissionable plutonium from irradiated nuclear fuel. In the early stage of commercial nuclear power, reprocessing was thought essential to supplying nuclear fuel. Federally sponsored breeder reactor development included research into advanced reprocessing technology.

Note that the *Reprocessing* refers to the chemical separation of fissionable uranium and plutonium from irradiated nuclear fuel.

The World War II-era Manhattan Project developed reprocessing technology in the effort to build the first atomic bomb. With the development of commercial nuclear power after the war, reprocessing was considered necessary because of a perceived scarcity of uranium. Breeder reactor technology, which transmutes nonfissionable uranium into fissionable plutonium and thus produces more fuel than consumed, was envisioned as a promising solution to extending the nuclear fuel supply [50].

Commercial reprocessing attempts, however, encountered technical, economic, and regulatory problems. In response to concern that reprocessing contributed to the proliferation of nuclear weapons, President Carter terminated federal support for commercial reprocessing. Reprocessing for defense purposes continued, however, until the Soviet Union's collapse brought an end to the Cold War and the production of nuclear weapons [50].

The DoE's latest initiative to promote new reactor technology using "proliferation-resistant" reprocessed fuel raises significant funding and policy issues for Congress. US policies that have authorized and discouraged nuclear reprocessing are summarized in reference [50].

Several commercial interests in reprocessing foundered due to economic, technical, and regulatory issues. President Carter terminated federal support for reprocessing in an attempt to limit the proliferation of nuclear weapons material. Reprocessing for nuclear weapons production ceased shortly after the Cold War ended. The DOE now proposes a new generation of "proliferation-resistant" reactor and reprocessing technology.

Global recognition of the need for such verification is reflected in the Treaty on the Non-Proliferation of Nuclear Weapons (the nonproliferation treaty [NPT]) [42]. Under the NPT, the nuclear weapons states (China, France, Russia, United Kingdom, and United States) undertake not to transfer nuclear weapons or any other nuclear explosive devices, and not to support manufacture or acquisition of such weapons or devices by any nonnuclear weapons states (NNWS). More than 180 NNWS are now party to the treaty, which means they have agreed that the International Atomic Energy Agency (IAEA) must apply safeguards on all their nuclear material—according to what are known as full-scope or comprehensive safeguards agreements [43].

Under the NPT, the Nuclear Weapon States (NWS) (i.e., China, France, Russia, United Kingdom, and United States) undertake not to transfer to any recipient whatsoever nuclear weapons or any other nuclear explosive devices or control over them, and not to support manufacture or acquisition of such weapons or devices by any NNWS [Article I].

NNWS party to the NPT undertake not to receive any nuclear weapons or other nuclear explosive devices, nor to accept assistance in this respect [Article II].

In summary, Safeguards are a set of technical measures applied by the IAEA on nuclear material and activities, through which the Agency seeks to independently verify that nuclear facilities are not misused, and nuclear material not diverted from peaceful uses. States accept these measures through the conclusion of safeguards agreements.

The IAEA safeguards are an essential component of the international security system. The NPT of Nuclear Weapons is the centerpiece of global efforts to prevent the further spread of nuclear weapons. Under the Treaty's Article 3, each NNWS is required to conclude a safeguards agreement with the IAEA.

In summary, Nuclear Non-Proliferation (NNP) is the effort to eliminate the spread of nuclear weapon technology, and to reduce existing stockpiles of nuclear weapons. Nuclear-weapon nations and their allies do not want any other nations or entities to get the technology, and those threatened by or otherwise hostile to these nations (or any other nations) often want to have it. Meanwhile, peaceful people across the world want no one to have nuclear weapons.

To reach their goal, people working for nonproliferation must secure and monitor existing nuclear weapons and weapon material, they must monitor facilities conceivably able to produce weapon material and minimize the construction of such facilities, and they must perform political gymnastics to minimize the allure of nuclear weapons to the world.

1.7 Nuclear Safeguards

Nuclear safeguards system was created to address the nuclear threat and preventing the spread of nuclear weapons built by wrong nations and terrorist acts. This idea was established in the late 1950s as precautionary responses to manufacturing and testing of nuclear weapons by several states, with the establishment of the IAEA and Euratom. IAEA Safeguards are a system of inspection and verification of the peaceful uses of nuclear materials as part of the NNP Treaty, supervised by the IAEA

Under the IAEA guidelines, nuclear safeguards are measures to verify that countries comply with their international obligations not to use nuclear materials for nuclear explosives.

Nuclear safeguards are measures to verify that countries comply with their international obligations not to use nuclear materials for nuclear explosives.

A fundamental principle of the safeguard's regime is that the verification is independent of the country and is performed by international inspectorates.

Alongside the IAEA safeguards requirements, civil nuclear material in the Member States of the European Union is also subject to the safeguard's provisions of Chapter VII of the Treaty establishing the European Atomic Energy Community [42] (the Euratom Treaty). The safeguards are applied by the European Commission to provide confidence that nuclear materials in the EU are not diverted from their declared end uses.

Bear in mind that, the existence of nuclear weapons is a continuous threat for the world. Considering the layout of MSR in Fig. 1.19, indicates that this reactor has been designed and configured from a nonproliferation and nuclear safeguard as well as safety point of view, so that it would fundamentally be considered a different class of nuclear reactor, which has been built and operated 3+time and the design:

1. **CANNOT** be used for nuclear weapons.
2. **Burns waste** from conventional reactors.
3. **CANNOT** meltdown or explode.

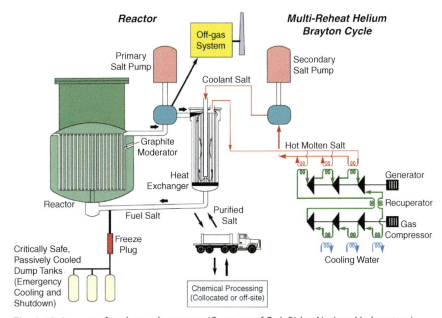

Fig. 1.19 Layout of molten salt reactor. *(Courtesy of Oak Ridge National Laboratory).*

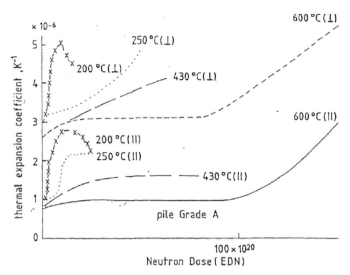

Fig. 1.20 MSR moderator curve. *(Courtesy of Oak Ridge National Laboratory).*

As far as a moderator in MSR is concerned, they are listed next. See Fig. 1.20 as well.
- Graphite exhibits complicated changes with irradiation, temperature, and stress that are not yet well understood.
  - Swelling/contraction.
  - Thermal expansion coefficient, and thus fuel tube dimensions
- MSRs require higher flux because fissile density is lower.
- Past graphite reactors were generally of a low power density.
  - MSRE, gas-cooled reactors
  - Four alternatives:
    - Overcome graphite challenges (Chinese approach).
    - Avoid graphite tubes (e.g., pebble bed).
    - Nonmoderated design—MSFR, Terrapower.
    - Adopt novel moderator material.

Moreover, MSR fuel cycles from point of fuel waste management is depicted in Fig. 1.21 and very much is self-explanatory.

In conclusion, we can state that, it is of fundamental importance for the international security to avoid new States acquiring nuclear weapons. This is the basis for the Nonproliferation Regime, whose main pillar is the NPT. International nuclear safeguards applied by the IAEA are the main tool of the Nonproliferation Regime.

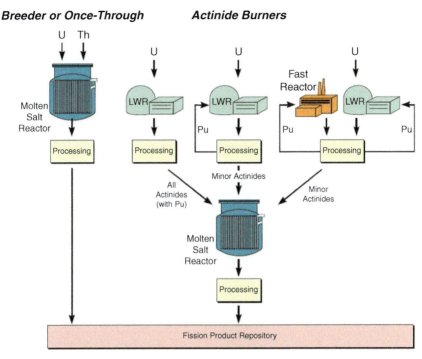

Fig. 1.21 Molten salt reactor fuel cycles. *(Courtesy of Oak Ridge National Laboratory).*

We can also state that:
- MSRs are Gen-IV concepts with potentially superior safety.
  - Concept examined in detail in the 1970s
  - Significant research required (particularly for actinide burning)
  - Modern participatory rural appraisal (PRA) techniques not yet applied
- Safety issues are significantly different.
  - Potential for major *reactor* accidents reduced
  - Potential for *processing* accidents increased
- Will require performance-based (not prescriptive) licensing strategy.

And most important, the MSR reactor has a small accident source term but a serious off-gas system

1.8 Safety by Physics Versus by Engineering

MSRs as we have said before are designed around the GEN-IV concepts that fall in category of high temperature AdvMSRs with potentially superior safety in mind [9].

As part of GEN-IV aspect of safety and nonproliferation, safety through national and international standards also should provide support for States with NPP capabilities in meeting their international obligations as well, thus will make safer when it comes to protection of these NPPs, both from radiation and proliferation point of view, while in production and providing the need for electricity within the grid [9].

One general international obligation for example is that a State must not pursue activities that cause damage in another State with or without NPPs on their soils, such as three major nuclear disasters that took place past decades up to including 2011 Japanese Fukushima Daiichi natural accident [9].

More specific obligations on contracting states are set out in international safety–related conventions. The international agreed that IAEA safety standards meet and provide the basic for States globally to demonstrate that they are meeting these obligations with few exceptions on these States that are not obeying the IAEA standards as being lonely nuclear power [9].

Traditional reactor safety systems are "active" in the sense that they involve electrical or mechanical operation on command. Some engineered systems operate passively, for example, pressure relief valves.

Both require parallel redundant systems. Inherent or full passive safety depends only on physical phenomena such as convection, gravity, or resistance to high temperatures, not on functioning of engineered components. Because small reactors have a higher surface area to volume (and core heat) ratio compared with large units, a lot of the engineering for safety (including heat removal in large reactors) is not needed in the small ones.

Comparison of current-generation NPP safety systems to potential SMRs designs is briefly listed in Table 1.2.

We can briefly mention safety-by-physics rather than safety-by-engineering and they are listed as follows:
- Fuel and coolant is the same, that is, loss of coolant = loss of power ⇒ **Cannot meltdown**.
- Atmospheric operating pressure ⇒ **No risk of pressure explosions**.
- No explosive gas production ⇒ **No risk of gas explosions**.
- Chemical stability in liquid ⇒ **No dangerous radioactive gasses**.
- Thermodynamic equilibrium control ⇒ **Walk-away safe**.

  More to be said about safety-by-engineering are listed below as well:
- Severity of accidents and system instability ⇒ **Safety-by-engineering**.
- Mainly innovation in safety ⇒ complexity ⇒ **Costs, risk, and delays**.
- Size becomes the economy of scale ⇒ **Poor market fit and massive upfront investments**.
- Larger reactors means increased severity ⇒ **Even more focus on safety**.

Table 1.2 Comparison of current generation nuclear power plant safety systems to potential SMR.

Current-generation safety-related systems	SMR safety systems
High-pressure injection system. Low-pressure injection system.	No active safety injection system required. Core cooling is maintained using passive systems.
Emergency sump and associated net positive suction head net pressure suction head (NPSH) requirements for safety-related pumps.	No safety-related pumps for accident mitigation; therefore, no need for sumps and protection of their suction supply.
Emergency diesel generators.	Passive design does not require emergency alternating current power to maintain core cooling. Core heat removed by heat transfer through vessel.
Active containment heat systems.	None required because of passive heat rejection out of containment.
Containment spray system.	Spray systems are not required to reduce steam pressure or to remove radioiodine from containment.
Emergency core cooling system initiation, Instrumentation, and Control (I&C) systems. Complex systems require significant amount of online testing that contributes to plant unreliability and challenges of safety systems with inadvertent initiations.	Simpler and/or passive safety systems require less testing and are not as prone to inadvertent initiation.
Emergency feedwater system, condensate storage tanks, and associated emergency cooling water supplies.	Ability to remove core heat without an emergency feedwater system is a significant safety enhancement.

Due to the inherent safety, MSRs are economically superior to conventional nuclear.

Furthermore, on issue of safety aspect of MSR, the following description would apply.

The inherent safety of MSRs is evident when trying to imagine any possible scenario for the release of radioactive material. As reviewed, within the salt itself there are virtually no volatile fission products as these are continuously removed during operation and stored well away from the reactor. Thus, while a salt spill is possible, fission products will remain within the salt. As well there are a full three levels of containment as with solid fueled reactors. In the molten salt case, the first barrier is the primary loop itself which is entirely contained within the second layer, a tight containment zone with only penetrations for pump drive shafts, intermediate coolant lines and possibly a shutdown rod. This containment zone that contains the primary heat exchangers is also made to collect and redirect any salt spills into decay heat dump tanks set up to deal with decay heat removal. This containment zone is also housed within an overall building containment to further assure no pathways for release [56].

The decay heat drain tanks also act in conjunction with passive freeze plugs to drain the salt to the safety of these tanks in any situation. If for any reason the core salt begins to rise in temperature, for example the failure of all pumps, this temperature increase melts a plug of frozen salt, which then drains the core salt to the dump tanks. After any use of the dump tanks, the salts can be pumped back up into the core for a restart of the reactor.

There is also absolutely no chemical or other driving potential for any major release. The system is not pressurized as the salts have very high (1400 °C) boiling points at ambient pressure. In fact, the primary salt loop is kept at the lowest pressure of the system so any leaks are inwards, the opposite of LWRs. There is no water used within containment that could lead to steam explosions or hydrogen production and subsequent detonation. No sodium or highly reactive substance that will react violently with water [56].

It should be mentioned that recent studies in France [57] on updated versions of the standard MSBR design showed a potential temperature reactivity problem which meant the original design might have actually had a slightly positive global temperature coefficient (the fast-acting term is indeed negative but a positive contribution occurring 10s of seconds later from the graphite heating leads to a net global positive coefficient). While there are many solutions for this for the pure thorium to ^{233}U cycle, a DMSR converter reactor has no such issue due to an enhancement of the negative terms from the presence of ^{238}U [56].

Any re-criticality events with the salts out of core are virtually impossible as they are only critical within the heterogeneous core graphite. Any external sabotage or explosions would render the core geometry and heterogeneity useless to reach criticality. See the next section.

1.9 Criticality Issue of Nuclear Energy Systems Driven by MSRs

MSRs, now often termed LFRs are no different than traditional reactor, when it comes to chain reaction and criticality issue.

A nuclear reactor works on the principle of a chain reaction. An initial neutron is absorbed by a fissile nuclide and during the process of fission; additional neutrons are released to replace the neutron that was consumed. If more neutrons are produced than are consumed, then the neutron population grows. If fewer neutrons are produced than are consumed, the neutron population shrinks. The number of fissions caused by the neutron population determines the energy released.

To quantify this concept, let us define a multiplication factor k. We will define k as the ratio of the production to consumption of neutrons [55].

$$k = \text{Multiplication Factor} = \frac{\text{Production}}{\text{Consumption}} \tag{1.1}$$

However, multiplication factor of a chain reactions takes place, when fissile nuclei as ^{233}U, ^{235}U, and ^{239}Pu, absorb, the excited compound nucleus breaks up into two fission products and releases several high-energy (~1 MeV) neutrons. When one of these neutrons is absorbed in another fissile nucleus, a chain reaction may begin.

In a hypothetical system of infinite size, neutrons do not leak out of the system. They are produced by fission and are removed only by absorption in the materials.

The infinite multiplication factor is defined as

$$k_\infty = \frac{\text{Neutrons produced in one generation}}{\text{Neutrons absorbed in the preceding generation}}$$
$$= \frac{\text{Rate of neutron production}}{\text{Rate of neutron absorption}} \tag{1.2}$$

The condition of criticality for stable self-sustaining fission chain reaction is that $k_\infty = 1$, in an infinite system without external neutron source.

In a system of finite size, some neutrons leak out of the system. The criticality condition must take into account the neutron leakage. The effective multiplication factor is given by:

$$k_{\mathit{eff}} = \frac{\text{Rate of neutron production}}{\text{Rate of neutron absorption} + \text{Neutron leakage}} \tag{1.3}$$

Using definition of multiplication factor k as above, then we can declare the chain reaction is time-independent, if $k = 1$, where the number of neutrons in any two consecutive fission generations are the same and at this point the reactor is in a critical mode. By the same talking if $k < 1$, then the reactor is in a subcritical condition and similarly if $k > 1$ the core is under a supercritical operating mode.

In summary, reactor core can operate under three following conditions:

$k < 1$	Subcritical
$k = 1$	Critical
$k > 1$	Supercritical

Having this knowledge in front of us, we can state that, the condition of criticality for stable self-sustaining fission chain reaction is that $k_\infty = 1$, in an infinite system without external neutron source.

In a system of finite size, some neutrons leak out of the system. The criticality condition must take into account the neutron leakage. The effective multiplication factor is given by Eq. (1.3).

The requirement for criticality in a finite system without external neutron source is that $k_{eff} = 1$.

If $k_{eff} < 1$, then the system is subcritical. In such a case the population of neutrons will steadily decrease. In the presence of an external neutron source, however, a steady state can be maintained by the source. When $k_{eff} > 1$, the system is supercritical.

In such a case the neutron production is larger than neutron loss, thus the population of neutrons will tend to increase exponentially with time. This increase could cause a nuclear explosion, or it could be limited by use of materials that absorb neutrons more strongly as they heat up, such as zirconium (or Uranium-238). More detail of calculation can be found in other study [55]

For purpose of reactor design, all the chosen materials and related subsystem within core (i.e., fuel rods) should be considered as part of criticality analysis so we can meet our primary objective of reactor design, which is having a reactor that can be in a critical mode. Sometimes we need reiterate our analysis and material that reactor finally goes critical as part of design and criticality condition is met, where $k_{eff} = 1$.

However, in terms of possible reactivity excursions MSRs are also superior. There is no excess reactivity needed during operation and no control rods that can be accidentally removed (some designs included low worth rods for minor temperature control). The salts have negative temperature reactivity coefficients dominated by Doppler shifting that act instantly. In transient studies even sudden (and difficult to imagine) reactivity insertions give prompt criticality, the salts merely jump up in temperature until they are subcritical again.

As with LWRs the reverse situation of sudden cooling must be planned for, but this is as simple as assuring pumps are such that a cold slug of salt cannot be moved into the core too quickly. As well, as the minimum salt temperature is already close to its freezing point, such an event is virtually impossible. In general, the high heat capacity of roughly 300 tons of salt also helps in smoothing any transient and makes any temperature rise from decay heating very gradual. See also Section 1.8 for safety issues.

1.10 Denatured Molten Salt Reactor (DMSR)

The general description of DMSR is described in this section. Historically, in 1980, the engineering technology division at ORNL published a paper entitled "Conceptual Design Characteristics of a Denatured Molten-Salt Reactor with Once-Through Fueling." In it, the scientist involved with the matter of DMSR "examine the conceptual feasibility of a molten-salt power reactor fueled with denatured uranium-235 (i.e., with LEU) and operated with a minimum of chemical processing." The main priority behind the design characteristics was proliferation resistance [12]. Although the DMSR can theoretically maybe fueled partially by thorium or plutonium, fueling solely with LEU helps maximize proliferation resistance.

Other important goals of the DMSR were to minimize R&D and to maximize feasibility. The GIF includes "salt processing" as a technology gap for MSRs [13]. The DMSR requires minimal chemical processing because it is a burner rather than a breeder. Both reactors built at ORNL were burner designs. In addition, the choices to use graphite for neutron moderation and enhanced Hastelloy-N for piping simplified the design and reduced R&D.

The plant concept is a direct outgrowth of the ORNL reference-design of MSBR (Fig. 1.22), and therefore it contains many favorable features of the breeder design and yet perfectly falls into category commercial NPP that obeys the NPT rules and regulations.

However, as we stated above, to comply with the antiproliferation goals, it also contains a number of differences, principally in the reactor core design and the fuel cycle.

Fig. 1.18 is a simple schematic diagram of the reference—design MSBR. At this level of detail, there is only one difference from DMSR concept and that is that the online chemical processing plant, which is shown at the left of the core, would not be required for the DMSR.

However, MSRs, now often termed LFR, come in many potential forms. All involve fluorides of fissile and fertile elements mixed within carrier salts that act as both fuel and coolant to transfers fission heat from a critical core to an intermediate heat exchanger.

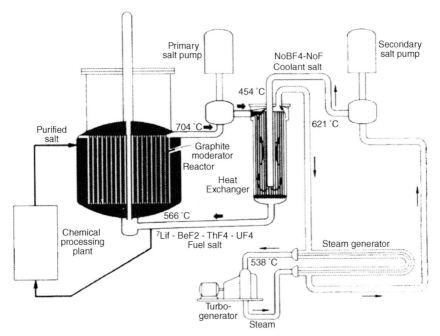

Fig. 1.22 Single-fluid, two-region molten salt breeder reactor.

There exists a broad range of design choices such as whether graphite is used as moderator or not, whether fuel processing for fission product removal is employed, whether the system runs in a denatured (LEU) state by the inclusion of ^{238}U and also whether one operates as a single-fluid or a two-fluid system (a two-fluid system has separate salts for fissile ^{233}U and fertile thorium).

Fig. 1.23 shows depiction of the 1950 graphite free on left-hand side, with two-region concept, and on right-hand side you observe, the 1960s intermixed two-fluid MSBR design using internal graphite Plumbing.

These choices also dictate whether a system has a breeding ratio (BR) > 1.0 (to produce excess fissile for future startups) or a BR = 1.0 to break even on fissile production or if BR < 1.0 making it a converter reactor requiring annual additions of fissile fuel of some kind.

As it was also indicated before in this chapter, a study was made to examine the conceptual feasibility of a molten-salt power reactor fueled with denatured ^{235}U and operated with a minimum of chemical processing.

Because such a reactor would not have a positive breeding gain, reductions in the fuel conversion ratio were allowed in the design to achieve other potentially favorable characteristics for the reactor. A conceptual core

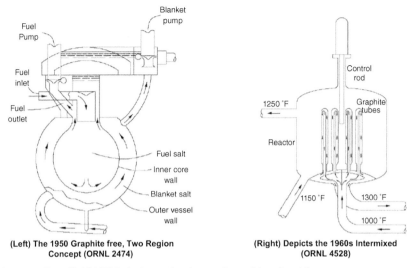

(Left) The 1950 Graphite free, Two Region Concept (ORNL 2474)

(Right) Depicts the 1960s Intermixed (ORNL 4528)

Fig. 1.23 Two-fluid MSBR design using internal graphite plumbing.

design was developed in which the power density was low enough to allow a 30-year life expectancy of the moderator graphite with a fluence limit of 3×10^{26} neutrons/m^2 (E > 50 Kev). Furthermore, we have learned that denatured MSR is derivative of MSR with idea of nonproliferation compliance in mind, solely with purpose of LEU that helps maximize proliferation resistance.

By now, we also know that MSRs are one of six next-generation designs chosen by the GEN IV program. Traditionally these reactors are thought of as thermal breeder reactors running on the thorium to ^{233}U cycle and the historical competitor to fast breeder reactors. However, simplified versions running as converter reactors without any fuel processing and consuming LEU are perhaps a more attractive option. Uranium consumption levels are less than 1/6th that of LWR or a 1/4th of canada deuterium uranium (CANDU) while at the same time offering clear advantages in safety, capital cost, and long-lived waste production along with increased proliferation resistance.

Again, keep in mind that MSRs are liquid-fueled reactors that can be used for burning actinides, producing electricity, producing hydrogen, and producing fissile fuels (breeding). Fissile, fertile, and fission products are dissolved in a high-temperature, molten fluoride salt with a very high boiling temperature (~1400 °C). The molten salt serves as both the reactor fuel and the coolant.

Heat is generated in the reactor core and transported by the fuel salt to heat exchangers [51] before returning to the reactor core.

The primary purpose of such study was to identify and characterize one or more DMSR concepts with antiproliferation attributes at least equivalent to those of a "conventional" LWR operating on a once-through fuel cycle. The systems were also required to show an improvement over the LWR in terms of fissile and fertile resource utilization.

Considerable effort was devoted *to* characterizing features of the concept(s) that would be expected to affect the assessment of their basic technological feasibility. These features included the estimated costs and time schedule for developing and deploying the reactors and their anticipated safety and environmental features.

Although the older MSR studies were directed toward a high-performance breeder [and a reference MSBR] [52] design [was developed], the basic concept is adaptable to a broad range of fuel cycles.

Aside from the breeder, these fuel cycles range from a plutonium burner ^{233}U production, through a DMSR with break-even breeding and complex on-site fission-product processing 99% to a denatured system with a 30-year fuel cycle that is once-through with respect to fission-product cleanup and fissile-material recycle. Of these, the past one currently appears to offer the *most* advantages for development as a proliferation-resistant power source. Consequently, this report is concentrated on a conceptual DMSR with a 30-year fuel cycle and no special chemical processing for fission-product removal.

The systems were also required to show an improvement over the LWR in terms of fissile and fertile resource utilization. Considerable effort was devoted *to* characterizing features of the concept(s) that would be expected to affect the assessment of their basic technological feasibility. These features included the estimated costs and time schedule for developing and deploying the reactors and their anticipated safety and environmental features.

More detailed about this particular reactor can be found in the study by Engel, J.R. (1980) [12].

1.11 MSR Pros and Cons

As we have said and learned so far, MSRs are nuclear reactors that use a fluid fuel in the form of very hot fluoride or chloride salt rather than the solid fuel used in most traditional Generation-III (GEN-III) or even some SMRs of GEN-IV. As the fuel salt is liquid, it can be both the fuel (producing the heat) and the coolant (transporting the heat to the power plant). There are many different types of MSRs, but the most talked about one is definitely the LFTR. This MSR has thorium and uranium dissolved in a fluoride salt and can get planet-scale amounts of energy out of our natural resources of thorium minerals, much like a fast breeder can get large amounts of energy out of our uranium minerals. There are also fast breeder fluoride MSRs that do not use thorium at all. And there are chloride salt–based fast MSRs that are usually studied as nuclear waste burners due to their extraordinary amount of very fast neutrons.

In this section, we list benefits (pros) and problems (cons) of MSRs in a brief format.

1. **Benefits of MSRs**

    The benefits of MSRs are plentiful, hence their resilience as an interesting topic throughout reactor history. We break them down by topic here.

    - **Sustainability**

        Sustainability is a measure of how efficiently a system can use natural resources. Most traditional reactors can only burn about 1% of the uranium on the Earth. Many advanced reactors, including MSRs, can do much better. Here is why MSRs are good in this regard.
        - *Online fission product removal*—As the fuel is liquid, it can be processed during operation. This means that when atoms split into the smaller atoms (fission products), those small atoms can be collected and pulled out of the core very quickly. This prevents those atoms from absorbing neutrons that would otherwise continue the chain reaction. This allows very high fuel efficiency in MSRs.
        - *Good utilization of thorium*—As mentioned above, the MSR chemical plant can continuously remove fission products and other actinides during operation. This means that when Thorium absorbs a neutron and becomes Pa-233, the Protactinium can be removed from the core and allowed to decay to U-233 in peace, without any risk of causing parasitic neutron losses. While this is not the only way to burn Thorium, it is perhaps the most elegant.
        - *No neutron losses in structure*—As there is no structure like cladding, fuel ducts, grid spacers, etc., there are no neutron losses in these. This helps fuel efficiency and therefore sustainability.

    - **Economics**

        Though economics are not truly known until a system is in commercial operation, there are reasons to think MSRs would have favorable low price.
        - *Online refueling*—Where normal reactors have to shut down to move fuel around or put new fuel in, MSRs can do all this while

at full power. You just dump in a new chunk of fuel into the vat (carefully, of course). This allows high-capacity factors, improving economics. The reactors may still have to shut down to do maintenance, but they likely will have better uptimes.
- *No fuel fabrication*—Any commercial fuel fabricator will tell you that it is expensive to build fuel assemblies, fuel pellets, cladding tubes, core support structures, flow orifices, etc. MSRs are basically just vats of fuel, so they are much simpler and cheaper in this regard.
- *High temperatures possible*—Molten salts were first looked at for their ability to go to very high temperatures. At high temperatures, power cycles convert heat to electricity with much less loss, giving you more money for a given amount of heat. Additionally, many industrial processes require high-grade heat, and these reactors could be used to that while producing electricity. Best of all, MSRs can work at high temperature without a pressurized coolant (as required in gas-cooled reactors).
- *Smaller containment*—As the system pressure is low and the heat capacity is high, the containments can be smaller and thinner.

- **Safety**

The most important aspect of a nuclear reactor is safety. Here is the good news for MSRs.
- *Very low excess reactivity*—As they can be continually refueled, there is no need to load extra fissile material to allow the reactor to operate for a long time. This means that it is difficult to have something happen (like an earthquake) that could cause a shift in geometry that inserts reactivity and causes a power spike.
- *Negative temperature coefficient of reactivity*—In general, if the fuel heats up, it expands and becomes less reactive, keeping things stable. Note that this is not always true in graphite-moderated MSRs.
- *Low pressure*—As the fuel and coolant are at atmospheric pressure, a leak in a tube does not automatically result in the expulsion of a bunch of fuel and coolant. This is a major safety advantage that enables passive decay heat removal (preventing things like what happened at Fukushima). The salts generally have extremely high heat capacity as well, so they can absorb a lot of heat themselves. On the other hand, their thermal conductivity is about 60× worse than liquid metal sodium.
- *No chemical reactivity with air or water*—The fuel salt is generally not violently reactive with the environment. So where LWRs

have hydrogen explosions and SFRs have sodium fires, MSRs do well. Of course, MSR leaks are still serious because it is not just coolant… it is extremely radioactive fuel.
- *Drain tank failure mechanism*—If something goes wrong in an MSR and the temperature starts going up, a freeze plug can melt, pouring the entire core into subcritical drain tanks that are intimately linked to an ultimate heat sink, keeping them cool. This is an interesting accident mitigation feature that is possible only in fluid fuel reactors.

2. **Problems with MSRs**

   All those wonderful benefits cannot possibly come without a slew of problems. Lots of people promote these reactors without acknowledging the issues, but not us! A reactor concept has to stand on its two feet even in the face of disadvantages (and we think the MSR can do this). Let us go through them
   - **Mobile fission products**

     The primary concern with MSRs is that the radioactive fission products can get everywhere. They are not in fuel pins surrounded by cladding, but are just in a big, sealed vat. You can put a double-layered containment around it, sure, but it is still challenging to keep them all accounted for. Where some of these fission products and actinides are radioactive, others have chemical effects that can eat away at the containment. The implications of this are many.
     - *Material Degradation*—With half the periodic table of the elements dissolved in salt and in contact with the containment vessel, there are lots of corrosion and related concerns. Noble metals will naturally plate out on cold metal surfaces. In a power reactor, a heat exchanger will be the coldest metal around, and so the heat transfer surfaces will need periodic replacement. At MSRE, tellurium caused cracking of the Hastelloy-N material. This was mitigated with chemistry, but similar problems may show up in long-lived power reactors.
     - *Tritium production*—If lithium is used in the salt, tritium will be produced, which is radioactive and extremely mobile (as it is small, it can go through metal like a hot knife through butter). ORNL used a special sodium fluoroborate intermediate salt to capture most of it, but a large amount still escaped to the environment.

- *Remote maintenance*—The chemical plants will need periodic maintenance, but all of the equipment will be highly radioactive. Expensive remote maintenance will be required. If graphite moderator is used, its replacement will also be remote and expensive.
- *Complex chemical plant*—Some of the fission product removal techniques are simple, such as the gas sparging to remove Xe and Kr, and noble metal plate out. But to do the more serious fission product (or actinide) separation, complex processes are required, such as the liquid Bismuth reductive process, volatilization, or electroplating. These have been studied in detail but are complex enough to be a disadvantage. Do not make us post a process flow diagram. (You can find one from the MSBR on page 8 of ORNL-TM-6413) [53].

- **Proliferation**

The main political barrier to MSRs is their perceived bomb-factory capabilities. If you talk to nonproliferation people [54] (i.e., See Section 1.6), they will tell you that as soon as the (solid) fuel pins are cut open, a technology is considered proliferative. The problem with MSRs, then, is that the fuel is already completely cut open and melted. You are halfway to a bomb already, they think. Here is what they are worried about.
- *Protactinium-233 decays to pure, weapons-grade U-233*—Many thorium-cycle MSRs have to capture Pa as it is produced, removing it from the system while it decays to U-233 and then reinserting it into the reactor. They have to do this because otherwise the Pa-233 absorbs too many neutrons to maintain a breeding cycle. The problem here is that that ex-core U-233 is basically pure weapons-grade U-233 which could be used to make a bomb. It usually comes with Zr, but separating Pa from Zr is simple. Not many common reactors require such a proliferative step in their fuel cycle. Many MSR concepts do not do this, but LFTRs require it. Therefore, the owner of an LFTR could be producing bombs on the side. Many of the ideas for mitigating this problem (such as U-232 contamination and denaturing) only help against diversion by a nefarious third party. The owners of the plant could side-step these kinds of fixes easily, and that is really what matters.
- *Inventory tracking is difficult*—Because a lot of materials plate out in the reactor and in the chemical plant, it is difficult to keep exact track of all of your actinides. The IAEA puts safeguards in reactors

to make sure that all the actinides are accounted for (to verify that no one's making bombs on the side) but it will be difficult for the IAEA to distinguish plate-out losses from actual proliferative losses.
- **Other minor issues**
There are a few other concerns, but these probably have practical solutions
    - *Unknown waste form*—It is not clear what nuclear waste from MSRs will look like. The salt itself is not contained enough to be put in a repository so someone will have to come up with a stable waste form.
    - *Electrical heaters are required to stay liquid*—In a prolonged power outage, the colder parts of the heat transfer loop might solidify. This could cause temperatures to rise over in the core (which will of course still be self-heated liquid).

3. **Exotics in the salt**
For MSRs to breed in a thermal spectrum, the lithium in the salt must be enriched to very pure Li-7. Li-6 is a strong neutron poison and becomes tritium, the pesky mobile radiation source. Also, Beryllium in Flibe is a controlled substance. It has some weapons applications and is a very dangerous material in terms of biological inhalation risks. In chloride salts, you must enrich to have pure Cl-37. Otherwise, Cl-35 has a strong (n, proton) threshold reaction that poisons the reactor. Also, the activation product Cl-36 is a long-lived, water-soluble, hard beta emitter that complicates waste disposal. These enrichment needs increase the cost and complexity of MSR fuel cycles. Maybe we can find a more ideal salt someday.

4. **Summary**
There are more specific problems with more specific types of MSRs, but you get the general idea here. Basically, MSRs are underdeveloped and require a lot more research (especially in corrosion) before they can surely take off as the world's fleet of power plants. Personally, we think (as did Alvin Weinberg and Edward Teller) that these reactors have a shining place in the mid-future. Right now, we have to keep studying them! Fortunately, much work is ongoing.

In summary we can list all known "*Pros*" and "*Cons*" known to MSR here as follows;

Pros:
- Carbon-free operation.
- Inherently far safer than conventional LWRs.

- Abundant fuel (thorium).
- Chemically stable.
- Currently being developed in China and by US companies like Flibe.
- Very small amount of low-level radioactive waste. Should be much easier to manage.
- Concentrated energy source, requiring far less land than solar.
- Runs round the clock, good base-load and load-following source.
- Less suitable for weapons proliferation that conventional nuclear.
- Relatively low cost and scalable.
- Could potentially be used in a distributed manner.
- Technology is currently at the demonstration phase.
- Requires less cooling water than conventional reactors.

Cons:
- Nonrenewable fuel.
- Still produces hazardous waste (though far less).
- Can still facilitate proliferation of nuclear weapons.
- Quite different than current technology.
- Primarily conceived as a centralized plant.
- Like all big plants, could be a terrorist target.
- Technology not ready for prime time yet.
- Competes with renewables for investment dollars.

Because of the first two items on the cons list, I consider thorium to be ultimately unsustainable in the very long term.

1.12 The Potential of the MSR Concept

The fuel is fluid at reactor temperature, thereby eliminating extra costs associated with fabrication, handling, and reprocessing of solid fuel elements. Burn-up in the fuel is not limited by radiation damage or reactivity loss. The fuel can be reprocessed continuously in a side stream for the removal of fission products, and new fissionable material can be added, while the reactor is in operation.

MSRs can operate at high temperature, thus is an excellent candidate for nuclear air combined cycle (NACC) by increasing thermal efficiency to a better output and consequently it will reduce the cost of producing electricity on demand as result of population growth. See references [9 and 53] as well as Appendices A, C, and D.

This type of reactors produce high pressure superheated steam to achieve thermal efficiencies in the heat-power cycle equal to the best fossil fuel plants. The relatively low vapor pressure of the salt permits use of low pressure containers and piping.

The negative temperature coefficient of the reactor and low excess reactivity are such that nuclear safety is not primarily dependent upon fast-acting nuclear fuel rods.

The fuel salt has a low cross-section for the parasitic absorption of neutrons. When it is used with bare graphite as the moderator, very good neutron economies can be achieved. Molten-salt reactors thus are attractive as highly efficient converters and breeders in the Thorium-Uranium fuel cycle.

The fluoride salt used as the fluid fuel mixture have good thermal and radiation stability and do not undergo violent chemical reactions with water or air. They are compatible with the graphite moderator and can be contained satisfactorily in a specially developed high-nickel alloy. The volumetric heat capacity, viscosity, thermal conductivity and other physical properties also are within desirable ranges.

Use of relatively high circulation rates and temperature differences result in high mean power density, high specific power, and low fuel inventory.

1.13 Conclusions

As we stated in this chapter that MSRs are one of six next-generation designs chosen by the GEN IV program.

Traditionally these reactors are thought of as thermal breeder reactors running on the thorium to ^{233}U cycle and the historical competitor to fast breeder reactors. However, simplified versions running as converter reactors without any fuel processing and consuming LEU are perhaps a more attractive option.

MSRs, now often termed LFRs, come in many potential forms. One of these forms is FHR that is also considered as GEN-IV series of chosen reactor for future of NPP. All involve fluorides of fissile and fertile elements mixed within carrier salts that act as both fuel and coolant to transfers fission heat from a critical core to an intermediate heat exchanger. There exists a broad range of design choices such as whether graphite is used as moderator or not, whether fuel processing for fission product removal is employed, whether the system runs in a denatured (i.e., LEU) state by the inclusion of ^{238}U.

MSRs are either operated as a single-fluid or a two-fluid system (a two-fluid system has separate salts for fissile ^{233}U and fertile thorium). These choices also dictate whether a system has a BR > 1.0 (to produce excess

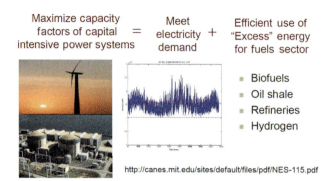

Fig. 1.24 Fluoride-salt-cooled high-temperature reactor (couples to hybrid nuclear renewable systems.

fissile for future startups) or a BR = 1.0 to break even on fissile production or if BR < 1.0 making it a converter reactor requiring annual additions of fissile fuel of some kind.

In conclusion, MSRs, can couple to Hybrid Nuclear Renewable Systems and they can be considered as based-load nuclear plant for variable electricity and process heat for even producing hydrogen. See Appendix D as well as the Fig. 1.24.

Also, in conclusion, we may say the traditional MSR with FHR can be considered as a backup as part of work and R&D on MSR implication for FHR.

In summary of this conclusion as illustrated in Fig. 1.25, MSR as one candidate of the Gen IV advanced nuclear power systems attracted more attention in China as it ranked top in fuel cycle and thorium utilization.

Two types of MSR concepts were studied and developed in parallel, namely the MSR with liquid fuel and that with solid fuel. Abundant fundamental research including the neutronics modeling, thermal-hydraulics modeling, safety analysis, material investigation, molten salts technologies etc. were carried out. Some analysis software such as COUPLE and FANCY were developed.

Several experimental facilities such as high-temperature fluoride salt experiment loop have been constructed. Some passive residual heat removal systems were designed, and one test facility is under construction. The key MSR techniques including the extraction and separation of molten salt and construction of N-base alloy have been mastered. Based on these fundamental research studies, the Chinese Academy of Sciences has completed the design of thorium-based MSRs with solid fuel and liquid fuel and is promoting their construction in the near future.

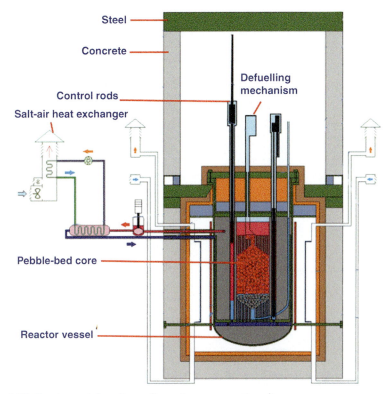

Fig. 1.25 Fundamental molten salt reactor cross-section diagram.

In China, future efforts should be paid to the material, online fuel processing, thorium–uranium fuel cycle, component design, and construction and thermal-hydraulic experiments for MSR, which are rather challenging nowadays.

References

[1] Emme, Eugene M, Aeronautics and Astronautics: An American Chronology of Science and Technology in the Exploration of Space, 1915–1960, LLC, Washington, DC, 1961, pp. 49–63.
[2] Megazone, The Decay of the Atomic Powered Aircraft Program, Worcester Polytechnic Institute, Worcester, Massachusetts, 1993 (Accessed 5 November 2008).
[3] Review of Manned Aircraft Nuclear Propulsion Program. Comptroller General of the United States. B-146759. 1963-02-28. 1963 https://fas.org/nuke/space/anp-gao1963.pdf (Accessed 10 January 2020).
[4] H.G. MacPherson, The molten salt reactor adventure, Nucl. Sci. Eng. 90 (1985) 374–380.
[5] H.G. MacPherson, et al., Summary report, fluid fuels reactor task force TID-8505, U.S. Atomic Energy Commission (1959).

[6] https://en.wikipedia.org/wiki/Molten_salt_reactor (Accessed 10 January 2020).
[7] F. Arkin, An Alternative Fuel for Nuclear Energy Looms, 2016 (Accessed 15 July 2016).
[8] Molten Salt Reactors. World Nuclear Association. https://www.world-nuclear.org/information-library/current-and-future-generation/molten-salt-reactors.aspx (Accessed 10/01/2020).
[9] B. Zohuri, PJ. McDaniel, Advanced Smaller Modular Reactors: An Innovative Approach to Nuclear Power, First ed., Springer Publishing Company, NY, 2019.
[10] Archived copy (PDF). Archived from the original (PDF) on, 2016-03-03. Retrieved 2014-01-21P.N. Haubenreich & J.R. Engel (1970). "Experience with the Molten-Salt Reactor Experiment" (PDF, reprint). Nuclear.
[11] P.N. Haubenreich, J.R. Engel, Experience with the molten-salt reactor experiment., Nucl. Appl. Technol. 8 (1970) 118–136.
[12] J.R. Engel, Conceptual Design Characteristics of a Denatured Molten-Salt Reactor with Once-Through Fueling., Oak Ridge National Laboratory: Engineering Technology Division, 1980.
[13] A Technology Roadmap for Generation IV Nuclear Energy Systems. U.S. DOE Nuclear Energy Research Advisory Committee and the Generation IV International Forum. https://www.gen-4.org/gif/jcms/c_40473/a-technology-roadmap-for-generation-iv-nuclear-energy-systems. 2002.
[14] *The Alvin Weinberg Foundation.* Archived from the original on 5 March 2016. https://en.wikipedia.org/wiki/Molten_salt_reactor (Accessed 10 January 2020).
[15] J. Smith, W.E. Simmons (Eds.), An Assessment of a 2500 MW$_e$ Molten Chloride Salt Fast Reactor, United Kingdom Atomic Energy Authority Reactor Group, http://www.egeneration.org/wp-content/Repository/Chloride_Salt_Fast_Reactor/AEEW-R956.pdf (Accessed 13 June 2015).
[16] W.C. May, W.E. Simmons (Eds.), Conceptual Design and Assessment of a Helium-cooled 2500 ME$_e$ Molten Salt Reactor With Integrated Gas Turbine Plant, United Kingdom Atomic Energy Authority Reactor Group. http://www.hellenicaworld.com/Science/Physics/en/MoltenSaltReactor.html (Accessed 13 June 2015).
[17] Ehresman, T. (ed.). *Molten Salt Reactor (MSR)(Fact Sheet). 08-GA50044-17-R1 R6-11.* Idaho National Laboratory http://www4vip.inl.gov/research/molten-salt-reactor/.
[18] *Kirk Sorensen has Started a Thorium Power Company Flibe Agile.* https://www.nextbigfuture.com/2011/05/kirk-sorensen-has-started-thorium-power.html, (Accessed 26 October 2011).
[19] Flibe Energy. flibe-energy.com.
[20] *Live chat: nuclear thorium technologist Kirk*, The Guardian, 2011 7 September.
[21] *New Huntsville company to build thorium-based nuclear reactors*, Archived 6 April 2012 at the Wayback Machine. huntsvillenewswire.com.
[22] *New nuke could power world until 2083*, The Register, 2013 14 March. https://www.theregister.com/2013/03/14/nuclear_reactor_salt/.
[23] Transatomic, Transatomic Power, Twitter. (25 September 2018). (Accessed 13 October 2019).
[24] Energy Department Announces New Investments in Advanced Nuclear Power Reactors…, US Department of Energy. (Accessed 16 January 2016).
[25] Integral Molten Salt Reactor. terrestrialenergy.com.
[26] Pre-Licensing Vendor Design Review, Canadian Nuclear Safety Commission. https://nuclearsafety.gc.ca/eng/reactors/power-plants/pre-licensing-vendor-design-review/ (Accessed 10 November 2017).
[27] Advances in Small Modular Reactor Technology Developments. International Atomic Energy Agency (IAEA). https://aris.iaea.org/Publications/SMR-Book_2018.pdf (Accessed 22 December 2019).
[28] Copenhagen Atomics - Thomas Jam Pedersen @ TEAC10. YouTube. 17 November 2019. (Accessed 22 December 2019).

[29] https://www.dualports.eu/wp-content/uploads/2019/06/Seaborg-making-nuclear-sustainable.pdf. *Missing or empty* |title= (help).
[30] http://seaborg.co/". External link in |title= (help).
[31] European Commission: CORDIS : Projects & Results Service: Periodic Report Summary – EVOL (Evaluation and viability of liquid fuel fast reactor system). Archived from the original on 13 April 2016.
[32] *EVOL (Project n 249696) Final Report.* https://cordis.europa.eu/docs/results/249/249696/final1-final-report-f.pdf.
[33] J. Serp, M. Allibert, O. Beneš, S. Delpech, O. Feynberg, V. Ghetta, D. Heuer, D. Holcomb, V. Ignatiev, The molten salt reactor (MSR) in generation IV: Overview and perspectives, Prog. Nucl. Energy 77 (2014) 308–319, doi:10.1016/j.pnucene.2014.02.014.
[34] SAMOFAR home. SAMOFAR. (Accessed 31 August 2018).
[35] J. Thomas, Dolan, Molten Salt Reactors and Thorium Energy. First Ed., Woodhead Publishing, Cambridge, MA, USA, 2017, pp. 557–564. ISBN 978-0-08-101126-3.
[36] Evans-Pritchard, Ambrose, China blazes trail for 'clean' nuclear power from thorium The Daily Telegraph, 2013 UK. https://www.reddit.com/r/energy/comments/163cos/china_blazes_trail_for_clean_nuclear_power_from/. (Accessed 18 March 2013).
[37] D. Clark, China enters race to develop nuclear energy from thorium, The Guardian, (16 February 2011).
[38] M. Halper*China Eyes Thorium MSRs for INDUSTRIAL Heat, Hydrogen; Revises Timeline.* Weinberg Next Nuclear. The Alvin Weinberg Foundation. (Accessed 9 June 2016).
[39] S. Chen, China Hopes Cold War Nuclear Energy Tech Will Power Warships, Drones, 2017 https://www.scmp.com/news/china/society/article/2122977/china-hopes-cold-war-nuclear-energy-tech-will-power-warships (Accessed 4 May 2018).
[40] *Research clarifies origin of superior properties of new materials for next-generation molten salt reactors - ANSTO.* ansto.gov.au.
[41] *Molten salt reactor research develops class of alloys.* https://www.world-nuclear-news.org/Articles/Molten-salt-reactor-research-develops-class-of-all.
[42] http://www.onr.org.uk/safeguards/glossary.htm#NPT.
[43] http://www.onr.org.uk/safeguards/what.htm.
[44] M. El-Genk. Space nuclear reactor power system concepts with static and dynamic energy conversion. Energy Covers. Manag. 49(3) (2008) 402-411.
[45] C. Weidemann, M. Oswald, S. Stabroth, H. Klinkrad, P. Vorsmann, Size distribution of NaK droplets released during RORSAT reactor core ejection, Adv. Space Res. 35 (2005) 1290–1295.
[46] M. Van Dyke, et al., The Safe Affordable Fission Engine (SAFE) Test Series, Proceedings of the NASA/JPL/MSFC/UAH 12th annual advanced space propulsion workshop, University of Alabama, *Alabama*, 2001.
[47] IAEA,. The role of nuclear power and nuclear propulsion in the peaceful exploration of space. https://www-pub.iaea.org/MTCD/publications/PDF/Pub1197_web.pdf.
[48] S. Wangab, C. Chaohui He, Subcritical space nuclear system without most movable control systems, J. Nucl. Sci. Tech. 52 (12) (2015).
[49] P. Yvon, F. Carre´, Structural materials challenges for advanced reactor systems, J. Nucl. Mater. 385 (2009) 217–222.
[50] Anthony Andrews, Nuclear Fuel Reprocessing: U.S. Policy Development, Congressional Research Service (CRS) The Library of Congress, Prepared for Members and Committees of Congress, Order Code RS22542,Updated March 27, 2008 https://www.everycrsreport.com/reports/RS22542.html (Accessed 10 January 2020).
[51] B. Zohuri, Compact Heat Exchangers: Selection, Application, Design and Evaluation., First ed, Springer Publication Company, New York, 2016.

[52] C.R Roy (Ed.), Conceptual Design Study of a Single-Fluid Molten-Salt Breeder Reactor, ONRL-4541, Oak Ridge National Laboratory, Oak Ridge, Tennessee, 1971.
[53] J.R. Engel, W.R. Grimes, W.A. Rhoades, J.F. Dearing, Molten-Salt Reactors for Efficient Nuclear Fuel Utilization Without Plutonium Separation, ORNL/TM-6413, Oak Ridge National Laboratory, Oak Ridge, Tennessee, 1978.
[54] https://whatisnuclear.com/non-proliferation.html (Accessed 10 January 2020).
[55] B. Zohuri, Neutronic Analysis for Nuclear Reactor Systems., Second ed, Springer Publishing Company, New York, 2019.
[56] D. LeBlanc, Denatured Molten Salt Reactors (DMSR): An Idea Whose Time Has Finally Come?, Physics Department, Carleton University, Ottawa, Canada, 2010.
[57] S. Thomas, The demise of the pebble bed modular reactor, Bull. At. Sci. (2009) June.
[58] H.W. Savage, Components of the Fused-Salt and Sodium Circuits of the Aircraft reactor Experiment, ORNL-2348. In: Metallurgy and Ceramics, TID-4500 (13th edition Rev.), 1958, https://www.osti.gov/servlets/purl/4308571 (Accessed 10 January 2020).
[59] Megazone, The Decay of the Atomic Powered Aircraft Program, Worcester Polytechnic Institute, Worcester, Massachusetts, 1993 http://www.islandone.org/Propulsion/AtomPlane.html (Accessed 5 November 2008).

CHAPTER 2

Integral Molten Salt Reactor

2.1 Introduction

Terrestrial Energy's (TE's) integral molten salt reactor (IMSR) has entered the second phase of a vendor design review by the Canadian Nuclear Safety Commission (CNSC). The design was the first advanced reactor to complete the first phase of the CNSC's regulatory prelicensing review. The illustration Fig. 2.1 is presentation of artistic facility layout of an molten salt reactor (MSR) at Oak Ridge National Laboratory (ORNL).

The CNSC's prelicensing vendor design review is an optional service to provide an assessment of a nuclear power plant (NPP) design based on a vendor's reactor technology. It is not a required part of the licensing process for a new NPP but aims to verify the acceptability of a design with respect to Canadian nuclear regulatory requirements and expectations.

The review involves three phases: a prelicensing assessment of compliance with regulatory requirements; an assessment of any potential fundamental barriers to licensing; and a follow-up phase allowing the vendor to respond to findings from the second phase. These findings will be taken into account in any subsequent construction license application, increasing the efficiency of technical reviews, according to the CNSC.

TE completed phase 1 of the vendor design review in November 2017. CNSC said the company had demonstrated an understanding of the regulator's requirements applicable to the design and safety analysis of the 400 MWt IMSR, known as IMSR400. Terrestrial Energy had also demonstrated its intent to comply with CNSC regulatory requirements and expectations for NPP, it said.

TE announced that phase 2 of the review has now begun. This, it said, involves a detailed follow-up of phase 1 activities, and an assessment of the IMSR design's ability to meet all 19 focus areas of power plant licensing.

"This is a critical commercial step that precedes site selection and construction of the first plant," it said. Phase 2 is expected to take 2 years to complete.

MSRs use fuel dissolved in a molten fluoride or chloride salt, which functions as both the reactor's fuel and its coolant. This means that such a reactor could not suffer from a loss of coolant leading to a meltdown.

Fig. 2.1 Molten salt reactor facility layout. *(Source: Oak Ridge National Laboratory).*

Terrestrial Energy's IMSR integrates the primary reactor components, including primary heat exchangers, to a secondary clean salt circuit, in a sealed and replaceable core vessel. It is designed as a modular reactor for factory fabrication and could be used for electricity production and industrial process heat generation. The company aims to commercialize the modular reactor design in the late 2020s [1].

In June 2017, TE began a feasibility study for the siting of the first commercial IMSR at Canadian Nuclear Laboratories' Chalk River site. In March this year, Terrestrial and US utility Energy Northwest agreed a memorandum of understanding on the terms of the possible siting, construction and operation of an IMSR at a site at the Idaho National Laboratory in southeastern Idaho.

Last month, TE USA announced that it has partnered with utility Southern Company and several US Department of Energy national laboratories to investigate the production of hydrogen using its IMSR. The 2-year research and development project will examine the efficiency, design, and economics of using the IMSR to produce carbon-free, industrial-scale hydrogen using the hybrid sulfur process [1].

John Barrett, president and CEO of the Canadian Nuclear Association, said: "Small modular reactors using innovative technologies are a great new opportunity for the nuclear industry. In the race to develop them, Terrestrial Energy's next-generation design is leading the way to commercial deployment with its innovative IMSR design. Completing CNSC's phase 2 design review will be a major milestone" [1].

2.2 Integral Molten Salt Reactor (IMSR) Descriptions

The IMSR is designed for the small modular reactor (SMR) market. It employs SMR technology which is being developed by the Canadian company TE [2]. It is based closely on the denatured molten salt reactor (DMSR), a reactor design from Oak Ridge National Laboratory (ORNL). It also incorporates elements found in the small modular advanced high-temperature reactor SmAHTR, a later design from the same laboratory. The IMSR belongs to the DMSR class of MSR and hence is a "burner" reactor that employs a liquid fuel rather than a conventional solid fuel; this liquid contains the nuclear fuel and also serves as primary coolant.

The Integral Molten Salt Reactor (IMSR) design is intended to be used for a variety of heat demand applications, ranging from power generation to cogeneration, or process-heat only. Fig. 2.2 is diagram of IMSR.

In 2016, TE engaged in a prelicensing design review for the IMSR with the Canadian Nuclear Safety Commission (CNSC) [3,4] and entered the second phase of this process in October 2018 after successfully completing the first stage in late 2017 [1,5]. The company claims it will have its first commercial IMSRs licensed and operating in the 2020s.

2.3 Integral Molten Salt Reactor (IMSR) Design

The IMSR is so-called because it integrates into a compact, sealed, and replaceable nuclear reactor unit, called the IMSR core-unit. The core unit comes in a single size designed to deliver 400 megawatts of thermal heat, which can be used for multiple applications. If used to generate electricity then the notional capacity is 190 megawatts electrical. The unit include all the primary components of the nuclear reactor that operate on the liquid molten fluoride salt fuel: moderator, primary heat exchangers, pumps, and shutdown rods [7]. For the control function, redundant, shutdown rods are also integrated into the IMSR core-unit.

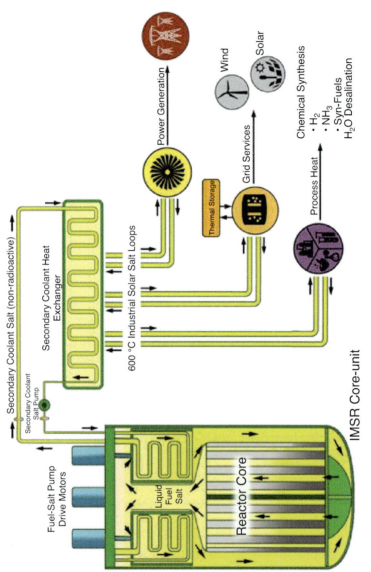

Fig. 2.2 Integral molten salt reactor layout. (*Source: www.wikipedia.org*).

Note that: These shutdown rods will shut down the reactor upon loss of forced circulation and will also insert upon loss of power.

The core unit forms the heart of the IMSR system. In the core unit, the fuel salt is circulated between the graphite core and heat exchangers. The core unit itself is placed inside a surrounding vessel called the guard vessel. The entire ore-unit module can be lifted out for replacement. The guard vessel that surrounds the core unit acts as a containment vessel. In turn, a shielded silo surrounds the guard vessel. Nominally for 7 years).

Note that: Since the design of the Integral Molten Salt Reactor (IMSR) is based on a small modular molten salt fueled reactor (e.g., IMSR400 with 400 MW Thermal), thus it features a completely sealed reactor vessel and very close to the image as illustrated in Fig. 2.3.

The sealed reactor vessel includes integrated pumps, heat exchangers, and shutdown rods as well.

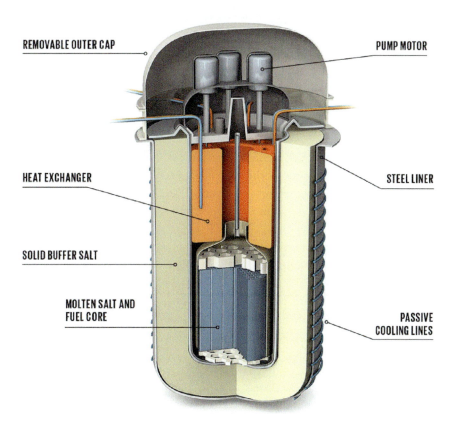

Fig. 2.3 Core unit and guard vessel in one of the two reactor silos. *(Source: IEEE Spectrum).*

MOLTEN SALT REACTOR

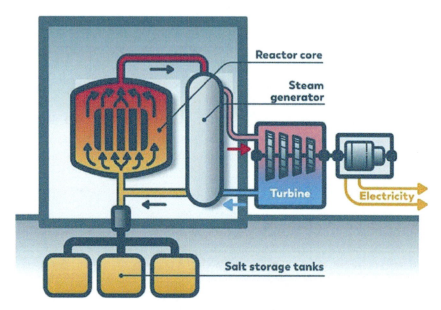

Fig. 2.4 IMSR400 main heat transport path for power generation, similar to MSR here. *(Source: Department of Energy).*

As we stated above, this vessel, called the IMSR "core unit," is replaced completely as a single unit at the end of its useful service life (nominally for 7 years). This allows factory production levels of quality control and economy, while avoiding any need to open and service the reactor vessel at the power plant site. Atypical diagram of IMSR 400 in close configuration as a MSR that is illustrated in Fig. 2.4.

Also, by now, we have learned that the IMSR uses molten fluoride salt, a highly stable, inert liquid with robust coolant properties, and high intrinsic radionuclide retention properties, for its primary fuel salt. A secondary, coolant salt loop, also using a fluoride salt (but without fuel), transfers heat away from the primary heat exchangers integrated inside the core unit. As shown in the figure above, the coolant salt loop, in turn, transfers its heat load to a solar salt loop, which is pumped out of the nuclear island to a separate building where it either heats steam generators that generate superheated steam for power generation or is used for process heat applications.

The IMSR belongs to the DMSR [8] class of MSR. It is designed to have all the safety features associated with the molten salt class of reactors including low pressure operation (the reactor and primary coolant is operated near normal atmospheric pressure), the inability to lose primary coolant (the fuel is the coolant), the inability to suffer a meltdown accident (the fuel operates in an already molten state) and the robust chemical binding of the fission products within the primary coolant salt (reduced pathway for accidental release of fission products).

The design is based heavily on the intensive molten salt reactor project of Oak Ridge National Laboratory (ORNL) in the 1950's–1970's. In this project, extensive research and development was done, and molten salt reactor materials and equipment were developed. This included such items as suitable graphite to act as neutron moderator, drain tanks for the salts, hot cells for housing the equipment, pumps, and heat exchangers. The ORNL program culminated in the construction and successful operation of the molten salt reactor experiment (MSRE). The MSRE employed a molten fluoride fuel salt, which was also the primary coolant. This fuel-coolant mixture was circulated between a critical graphite moderated core and external heat exchangers. The IMSR is essentially a reconfigured, scaled-up version of the MSRE. This minimizes required research and development. The IMSR has also been influenced by a more recent ORNL design, SmAHTR, which was to use solid fuel with fluoride salt coolant but had an "integral" architecture where all primary pumps, heat exchangers and control rods are integrated inside the sealed reactor vessel.

SmAHTR's cartridge core enables filling of the interassembly volume with nuclear-grade graphite, resulting in improved neutron utilization, as shown in Fig. 2.5. The entire core will be lifted out as a single unit for refueling. Each fuel assembly includes a molybdenum hafnium carbide control blade. While plate-style TRISO fuel bodies have not been manufactured previously, the manufacturing process steps of any geometric-shape fuel body are nearly identical.

IMSR is a proprietary design in which all the primary components [i.e., pumps, moderator, and primary heat exchanger (HX)] of the reactor core are sealed in a compact and replaceable component, the IMSR core unit, which is a sealed and replaceable reactor and that is why it is labeled as IMSR. TE is targeting 2024 date to have an operational demonstration commercial reactor in place.

The IMSR uses molten fluoride salt, a highly stable, inert liquid with robust coolant properties and high intrinsic radionuclide retention properties, for its primary fuel salt. A secondary, coolant salt loop, also using a fluoride salt (but without fuel), transfers heat away from the primary heat

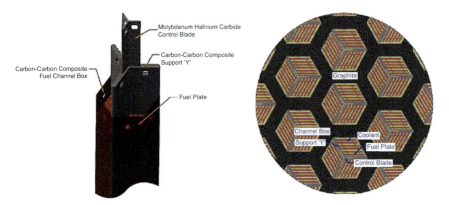

Fig. 2.5 SmAHTR fuel assembly upper end and partial core-cross section. *(Source: Oak Ridge National Laboratory).*

exchangers [6] integrated inside the core unit. The coolant salt loop, in turn, transfers its heat load to a solar salt loop, which is pumped out of the nuclear island to a separate building where it either heats steam generators that generate superheated steam for power generation or is used for process heat applications. The safety philosophy behind the IMSR is to produce a NPP with generation IV reactor levels of safety. For ultimate safety, there is no dependence on operator intervention, powered mechanical components, coolant injection, or their support systems, such as electricity supply or instrument air in dealing with upset conditions. This is achieved through a combination of design features: the inert, stable properties of the salt; an inherently stable nuclear core; fully passive backup core and containment cooling systems; and an integral reactor architecture.

The design uses standard assay low enriched uranium (LEU) fuel, with less than 5% U^{235} with a simple converter (also known as a "burner") fuel cycle objective (as do most operating power reactors today). The proposed fuel is in the form of uranium tetrafluoride (UF_4) blended with carrier salts. These salts are also fluorides, such as lithium fluoride (LiF), sodium fluoride (NaF), and/or beryllium fluoride (BeF_2). These carrier salts increase the heat capacity of the fuel and lower the fuel's melting point. The fuel salt blend also acts as the primary coolant for the reactor.

The DMSR, that is shown in Fig. 2.6 and it is carried into the IMSR design, proposed to use molten salt fuel and graphite moderator in a simplified converter design using LEU , with periodic additions of LEU fuel. Most previous proposals for molten salt reactors all bred more fuel than needed to operate, so were called breeders. Converter or "burner" reactors

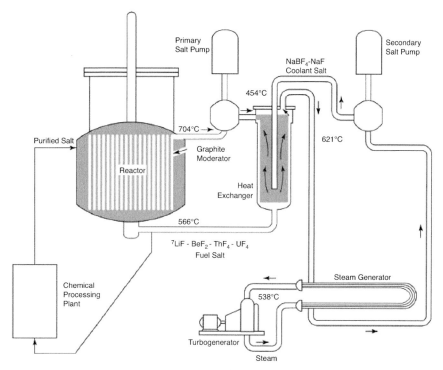

Fig. 2.6 The 1970s single fluid, graphite moderated molten salt breeder reactor. 1000 Mwe with a specific fissile inventory of 1500 kg. *(Reproduced from ORNL 4812).*

like the IMSR and DMSR can also utilize plutonium from existing spent fuel as their makeup fuel source. The more recent SmAHTR proposal was for a small, modular, molten salt cooled but solid TRi-structural ISOtropic fueled reactor [9].

The IMSR is a "thermal-neutron reactor" moderated by vertical graphite tubular elements. The molten salt fuel-coolant mixture flows upward through these tubular elements where it goes critical.

Note that: A thermal-neutron reactor is a nuclear reactor that uses slow or thermal neutrons. ("Thermal" does not mean hot in an absolute sense but means in thermal equilibrium with the medium it is interacting with, the reactor's fuel, moderator, and structure, which is much lower energy than the fast neutrons initially produced by fission) [10].

After heating up in this moderated core, the liquid fuel flows upward through a central common chimney and is then pulled downward by pumps through heat exchanges positioned inside the reactor vessel. The liquid fuel then flows down the outer edge of the reactor core to repeat the cycle.

All the primary components, heat exchangers, pumps, etc. are positioned inside the reactor vessel. The reactor's integrated architecture avoids the use of external piping for the fuel that could leak or break.

The piping external to the reactor vessel contain two additional salt loops in series: a secondary, nonradioactive coolant salt, followed by another (third) coolant salt. These salt loops act as additional barriers to any radionuclides and improve the system's heat capacity. It also allows easier integration with the heat sink end of the plant; either process heat or power applications using standard industrial grade steam turbine plants are envisioned by TE.

As we have also stated before, The IMSR core unit is designed to be completely replaced after a 7-year period of operation. This ensures that a sufficient operational lifetime of materials used in the IMSR reactor core can be achieved. During operation, small fresh fuel/salt batches are periodically added to the reactor system. This online refueling process does not require the mechanical refueling machinery required for solid fuel reactor systems.

The IMSR facility in cutaway bird's view is illustrated in Fig. 2.6. New modules are brought in by road (left) and are then lifted into the reactor cavity (middle right) by gantry crane. Also shown are secondary heat exchangers and manifolds that send heated molten salt to the power generating part of the plant (right, power generating building not shown).

2.4 Integral Molten Salt Safety Philosophy

The safety philosophy behind the IMSR is to produce a NPP with GEN-IV reactor levels of safety. For ultimate safety, there is no dependence on operator intervention, powered mechanical components, coolant injection or their support systems such as electricity supply or instrument air in dealing with upset conditions. This is achieved through a combination of design features:
- The inert, stable properties of the salt
- An inherently stable nuclear core
- Fully passive backup core and containment cooling systems.
- An integral reactor architecture

Overall safety of NPP falls into three fundamental safety requirements as follows:

1. *Control:* Failure of control systems or reactivity insertion events only leads to reactor stabilization at a slightly higher temperature.

2. *Cooling:* Inherent heat sinks are available initially to absorb transient and decay heat, with heat losses providing long-term cooling for both core and containment.
3. *Containment:* The fluoride salts are chemically stable, bind radioactive fission products to the salt, and have extremely high boiling points. Multiple engineered barriers are provided as backup to this inherent containment. Partial entrenchment of the reactor, combined with thick concrete and steel shielding, provides resistance to external events such as earthquakes, explosions, and aircraft crash.

An important part of the IMSR safety philosophy is to start by removing drivers that push radioactive material into the environment. Specifically, the reactor always operates at low pressure due to the inert, low-volatile fuel-coolant mixture and the absence of water or steam in the reactor. As the materials employed in the reactor system and even the interfacing systems, are all chemically compatible with each other, there is no potential for adverse chemical reactions such as fuel cladding hydrogen production or sodium reactions with air or water. This approach completely eliminates stored energy, both physical and chemical, from the reactor system. The IMSR further augments this high level of inherent, physics-based safety with its integrated, pipe-less, fail-safe systems architecture. The result is a simple and robust system with inherent safety.

As we know from our knowledge of light water reactor (LWR) of third generation (GEN-III), the safety concept of such reactor is totally depending on providing sufficient coolant flow to the solid fuel assemblies at all times. This must occur at high pressure. During loss of coolant accidents (LOCA) such as pipe leaks or breaks, reliable depressurization is required, followed by low-pressure coolant injection. These systems require various mechanical, electrical/control systems, instrumentation/sensing, instrument air and other support systems to operate reliably.

Note that: LOCA is a mode of failure for a nuclear reactor; if not managed effectively, the results of a LOCA could result in reactor core damage. However, due to radioactive decay, the nuclear fuel will continue to generate a significant amount of heat.

The IMSR does not depend on depressurizing the reactor or bringing coolant to the reactor. All required control and heat sink functions are already present where they are needed – in and directly around the IMSR core unit. As such, the IMSR completely eliminates any dependence on support systems, valves, pumps, and operator actions. This is the case in both the short term and the long term. To make this possible, the IMSR designers have combined molten salt reactor technology with integral reactor design and a unique cooling system.

Reactor criticality control is assured through a negative temperature feedback made possible by molten salt fuel. This negative temperature feedback assures reactor safety on overheating, even with loss of all control systems. Molten salt fuel does not degrade by heat or radiation, so there are effectively no reactor power limits to the fuel.

The unique cooling system is based on heat capacity and heat loss—which are immutable. Heat capacity is due to the thermal mass of the fuel salt, vessel metal, and graphite. Heat loss occurs as the reactor vessel is not insulated. Short-term cooling is provided by the low-power density core unit, and the internal natural circulation capability of the fluoride fuel salt, resulting in a large capacity to absorb transient and decay heat generation. Longer term cooling is provided by heat loss from the uninsulated reactor vessel which itself is enveloped by a guard vessel. This guard vessel is a closed vessel that envelops the core unit, providing containment and cooling through its vessel wall. Overheating of the core unit will cause the it to heat up and increase heat loss from the core unit, in turn increasing heat transfer, via thermal radiation, to the guard vessel. The guard vessel in turn is surrounded by a robust air cooling jacket. This cooling jacket will provide long term cooling. The cooling jacket operates at atmospheric pressure so will continue to cool in the event of a leak or damage to the jacket.

The guard vessel containment is provided as an additional hermetic barrier in the extremely unlikely event that the integral core unit itself would experience major failures. Without sources of pressure in the core unit (See Fig. 2.3) or in the containment itself, the containment is never challenged by pressure. Overheating of the containment is precluded by the balance of heat losses and heat generation even if the core unit would fail. The containment itself is covered by thick horizontal steel radiation shield plates at the top. These plates also provide protection against extreme external events such as aircraft crash or explosion pressure waves and provide an additional heat sink in any overheating scenario.

Fig. 2.3 is also a presentation of the Core unit which loses of heat to an enveloping guard vessel, which in turn loses heat to a passive air cooling jacket. This provides backup cooling in the unlikely event all of the redundant normal heat exchangers are unavailable for any reason.

The IMSR has a highly attractive seismic profile due to its compact, integral (pipe-less) primary system. In addition, the below grade silo housing the core unit, and low profile buildings, result in a very low center of gravity. Apart from the reactor system and buildings themselves, earthquakes can often threaten support systems such as main power, backup power, battery power, instrumentation and control systems, emergency coolant injection lines, or pneumatic and hydraulic systems. As the IMSR does not rely on any of these support systems, or even the cooling jacket for ultimate safety, the IMSR's safety is inherently insensitive to earthquakes.

The IMSR system is a high temperature reactor system whose key safety systems are designed for extreme normal and emergency temperature profiles. No support systems are required for safe shutdown. As a result of these features, the design is inherently insensitive to fires from a safety viewpoint.

As explained above, ultimate safety is provided by inherent and fail-safe features. Nevertheless, with a view of further improving the robustness of the design, to improve the reliability, and as investment protection, a considerable amount of defense-in-depth is built-in the systems. For the control function, redundant, shutdown rods are also integrated into the IMSR core unit. These shutdown rods will shut down the reactor upon loss of forced circulation and will also insert upon loss of power. Another backup is provided in the form of melt-able cans, filled with a liquid neutron absorbing material that will shut down the reactor on overheating. A complete loss of flow is however itself unlikely, as redundant primary pumps are used for circulating the fuel salt, so that the system can continue to operate at full power with any single pump failed or tripped, and slightly de-rated operation is possible with any two pumps failed or tripped. Similarly, to drive the pumps, conventional backup engine-driven power is available upon loss of main power. For containment, in addition to the inherently stable salt properties that act as physical and chemical containment, the integrated reactor architecture maximizes the integrity of the primary reactor system, the core unit. This makes leaks very unlikely. Should any leaks occur, a conventional, leak-tight containment is also provided. Overpressure failure is not a plausible event due the very high boiling points of the salts and to a lack of pressurization sources inside the core unit and even in interfacing systems. Overheating failure of the core unit is precluded by the use of high quality, high temperature capable materials in conjunction with the inherent heat loss to the surrounding passive cooling jacket.

2.5 Proliferation Defense

The IMSR has intrinsic, design, and procedural features that resist proliferation of fissile materials for weapons production:
1. Fuel cycle: the IMSR core unit itself is fueled with LEU on a once-through fuel cycle without fuel processing. There is no need for highly enriched uranium, and, with no fuel processing on-site, there is no possibility of separating fissile materials from the bulk of fuel salt. In addition, the high burn-up and lack of online fuel removal from the IMSR Core unit results in extremely poor isotopic quality in the spent fuel inventory. In a molten salt reactor such as the IMSR, fuel and coolant are intrinsically mixed. Removal of any significant amount of fuel therefore entails the removal of much larger volumes of coolant salt. Any future fuel processing would be done in a secure, centralized facility.
2. Systems design: the plant engineering does not allow the removal of significant quantities of fuel salt to small, and thus mobile shielded containers during normal operation. Fuel is added to the core unit, fuel salt need not be removed during the 7-year cycle. Minor transfers for chemistry monitoring and control or for final spent fuel salt transfer to large, shielded holding tanks will be completely human-inaccessible due to the high temperature and radiation levels. Any future transfer of spent fuel salt for sequestration or processing will involve time-consuming protocols under full supervision. Such off-site transfers would entail full self-protection due to activity levels of contained fission products.
3. Procedural controls: the IMSR will comply with all International Atomic Energy Agency proliferation safeguards, procedural controls, monitoring, and other requirements.

2.6 Safety and Security (Physical Protection)

The IMSR is a molten salt reactor, having a highly radioactive primary fuel-coolant loop. In addition, the high operating temperature means special shielded, insulated cells and systems will be employed. These features make the reactor design highly inaccessible to potential sabotage or terrorist attack. The IMSR also employs the integral architecture, which produces a compact, robust, and fully sealed primary core-heat exchanger unit. Another design feature that protects against external threats is the provision of a hardened below grade silo as the holding cell for the core unit. Such hardening and below grade construction provide superior protection against external events. The core unit itself is provided with additional layers of protection: a sealed guard vessel surrounding the bottom and sides of the Core unit, and a steel containment plate at the top. This containment head plate is made of very thick steel primarily for radiation shielding, which also makes for a high strength, ductile external events shield capable of stopping the impact of a large commercial aircraft crash or explosive devices. Because the IMSR does not rely on any support systems, any attack on such support systems does not impede any of the three primary safety functions—control, cooling, and containment—of the IMSR.

2.7 Description of Turbine-Generator Systems

The turbine-generator system does not serve any safety related mission and is not needed for ultimate safe shutdown of the IMSR core unit. Therefore, the IMSR will employ a standard, off-the-shelf, industrial grade superheated steam turbine plant. The use of a standard turbine avoids the necessity of having to develop a new turbine and control system for the IMSR and allows the highly flexible and reliable industrial grade steam turbines to be used, that can be sourced from a variety of major suppliers. These steam turbines are highly flexible, being capable of high degrees of bleed off steam for cogeneration/industrial heat purposes, whilst also being capable of operating as a condensing (power generating) turbine—only if required. This affords the IMSR access to many industrial power and heat markets such as steam assisted gravity drainage (SAGD), chemicals production, paper and pulp production, desalination, etc. In fact, these turbines, powered by fossil fuel boilers, are widely used already in many such industries today. The turbine-generator is a simple and compact, high speed, skid mounted design. A gearbox couples the high-speed turbine to a lower speed, 1500–1800 revolution per minute (RPM) electrical generator.

2.8 Electrical and Integrated and Circuit (I&C) Systems

The IMSR does not depend on electrical or even I&C systems for ultimate safety. For operability and investment protection, a few additional requirements do exist, such as freeze protection, as the fluoride salts have high freezing points. The normal plant operations systems provide this protection, in the form of pump trips, electrical (trace) heating, and thermocouple sensors. As with the steam turbine-generator, this type of equipment is available off-the-shelf from many suppliers. The pumps are the primary control units, as passive, flow-driven control rods are utilized. This links the trip and protection logic to the pump rather than to a control rod drive and logic system. The result is a simple and easy to control system where the only real variable to control actively is the pump speed (trace heating is typically kept in automatic mode for normal operations). For example, to rapidly shut down the reactor, no control rod drives, drive power, hydraulic supply, or other support systems are needed; instead, the pumps are simply tripped. The simplicity of this approach and the lack of dependence on any electrical systems also means that cyber-security is easy to address.

As the IMSR does not depend on the electrical systems for ultimate safety, the design of a detailed plant electrical and control room system is eased. However, as with the steam turbine-generator, standard, off-the-shelf industrial equipment is expected to be used due to the generic nature of the requirements.

2.9 Spent Fuel and Waste Management

The IMSR utilizes nuclear fuel with far greater efficiency than LWRs. There are three basic factors that improve the fuel utilization of the IMSR compared to LWRs:

1. The thermal-to-electrical efficiency is high; despite the modest size, the IMSR400 features higher plant electrical efficiency than LWRs. Depending on heat sink conditions and balance-of-plant arrangement, the IMSR400 generates 185 to 192 MWe from 400 MWth, resulting in a net efficiency of some 46–48%.
2. The fuel burn-up is high: rather than removing fuel from the reactor, new fuel is simply added to the total fuel charge forming the reactor core. The high burn-up and retention of fuel inside the core results in improved plutonium and minor actinide burning capability.
3. Without fuel cladding, internal metallic structures or H_2O, and with the IMSR design permitting the passive exit of Xenon from the core, there are far fewer parasitic neutron captures.

An additional advantage of the design is that its spent liquid fuel inventory is much easier to recycle than solid fuel elements, making it more attractive to recycle the fuel. As the fuel does not degrade, it can potentially be recycled many times. If fuel recycling is employed, Terrestrial Energy envisages this to occur in a central fuel recycling center, servicing many IMSR power plants.

The design also makes efficient use of engineering and construction materials. This is mostly due to the low operating pressure and lack of stored energy in the salts, eliminating the need for massive pressure vessels, pipe restraints, and containment buildings. High pressures are confined to the steam turbine plant, which is a very compact, standard industrial grade unit. The salts also have high volumetric heat capacity resulting in compact heat exchangers, pumps, and other heat transport equipment. The result is very compact equipment and small buildings.

The IMSR features a fully sealed unit, avoiding the need to open the reactor vessel during operation. This feature greatly reduces expected dose rates to personnel and avoids generation of large amounts of contaminated drainage water and disposables.

2.10 Plant Layout

The IMSR core unit is housed in a below grade silo. This forms the heart of the IMSR system. Surrounding the core unit are the guard vessel cooling jacket, and at the top, a removable steel containment head plate. All these components are located below grade as well. Above the containment head, shield plates cover the core unit for radiation shielding. An idle silo is provided next to the active silo, so that this silo can be loaded with a new core unit while facilitating cool down of a spent core unit. These two silos make up the main components of the reactor building. After draining the spent core unit followed by a cool down period, the empty spent module is the lifted out and stored in long-term storage silos in an area adjacent to the reactor building. Inlets and outlets are provided for the (nonradioactive) coolant salt lines coming into and out of the core unit, which transfer the heat to steam generators and re-heaters. The coolant salt lines are located inside the reactor building. The steam and reheat steam are utilized in a conventional, off-the-shelf turbine-generator system located in an adjacent building.

IMSR400 is a 400 MW thermal reactor that is suggested by Terrestrial Engineering (TE) of Canadian Company for commercial use.

2.11 Plant Performance

The IMSR plant is designed to accommodate various load users, from baseload to load-following. Featuring a simple modular and replaceable core unit, very high reliability ratings are targeted. The core unit has only a few moving parts (pumps) and these are redundantly fitted. Even premature replacement of a faulty module only mildly affects reliability and levelized cost. The use of an idling silo greatly increases plant capacity factor, as long cool down times (for radiation dose reduction) are possible without plant downtime.

The IMSR has been specifically designed for factory fabrication. Nuclear components are small and road-transportable. The IMSR core unit is designed for a short service lifetime, which allows dedicated factory lines to produce the units semiautomatically, similar to aircraft jet engine production lines for example.

The IMSR reactor building has a low profile and low mass, allowing rapid construction. The reactor building is a simple lightweight industrial building as it does not serve any major safety-related function - these are all provided by dedicated, robust, fail-safe systems.

Being a MSR, the IMSR has low fuel reload costs; fuel fabrication costs are zero, only extensive purification of the salts is required which, being bulk chemical processing, has a much lower cost than traditional high precision fuel fabrication with all its quality and process control costs. Fuel recycling costs are also much lower with molten salt fuels, as costly fuel deconstruction and – reconstruction steps are avoided and also because simple, compact, low waste volume creating distillation and fluorination processes can be utilized. This makes it likely that spent IMSR fuel will be recycled.

2.12 Development Status of Technologies Relevant to the Nuclear Power Plant

The IMSR is a simplified burner-type MSR. All of the basic technology has been proven during the operation of the MSRE at ORNL. The MSRE was a fully functional burner type MSR. Heat exchangers, pumps, suitable graphite, fuel salts and— purification steps, off-gas systems, fuel addition mechanisms, as well as alloys of construction were all successfully developed. The IMSR builds on this rich experimental evidence to assure a high degree of technological maturity. The IMSR utilizes an off-the-shelf industrial grade steam turbine-generator. Molten salt heated steam generators are also proven technology, with recent and very successful experience in advanced, large scale, salt-cooled concentrated solar power (CSP) projects. Similarly, water-cooled silo liners have been successfully employed in many nuclear reactor systems. Other technologies, such as tritium capture and management, are also proven technologies in the nuclear industry with a wide and current experience base in heavy water reactors as well as high temperature reactors. There are only a few components that have not been used before, such as the air cooling jacket. These components are technologically simple, and their proper function will be verified in large scale nonnuclear testing.

2.13 Development Status and Planned Schedule

The IMSR400 phase of work involves the support of a growing number of universities, third party laboratories and industrial partners. In addition, the Canadian nuclear regulator, the Canadian Nuclear Safety Commission (CNSC), at the request of Terrestrial Energy Incorporation (TEI), has started Phase 1 of a prelicensing Vendor Design Review of the IMSR. This is an assessment of a nuclear power plant design based on a vendor's reactor technology. The assessment is completed by the CNSC, at the request of the vendor. The words "prelicensing" signifies that a design review is undertaken prior to the submission of a license application to the CNSC by an applicant seeking to build and operate a new nuclear power plant. This review does not certify a reactor design or involve the issuance of a license under the *Nuclear Safety and Control Act*, and it is not required as part of the licensing process for a new nuclear power plant.

The basic engineering phase will be followed by the construction and start-up of a first full-scale commercial IMSR NPP at a site in Canada after securing all necessary operating licenses from the CNSC (and licenses from other authorities), a process that can be completed by early in the next decade.

2.14 Coupling IMSR Technology with Hybrid Nuclear/Renewable Energy Systems

The IMSR represents a clean energy alternative to fossil fuel combustion for industrial heat and provision, which is compact, efficient, and cost-competitive with fossil fuels. The IMSR is a Gen 4 reactor and a successor of the very effective MSRE work of ORNL (see Fig. 2.7).

The MSRE as shown in Fig. 2.6 operated from 1965 until 1969 using a fuel/moderator mixture of lithium, beryllium, thorium, and uranium fluorides [15]. It built on the success of the 1954 Aircraft Reactor Experiment [16], and on other early liquid-fuel schemes; the basic feasibility of such designs is well demonstrated. The graphite moderator ensures the correct

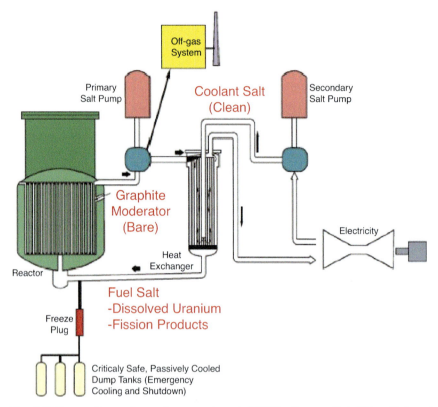

Fig. 2.7 The molten salt reactor experiment (MSRE). *(Source: Oak Ridge National Laboratory).*

neutron spectrum while the fuel salt circulates in a closed loop. Increased power (in a fluctuation) expands the liquid fuel, in turn reducing the power output. This safety feature was tested during MSRE operation, performing as expected. MSRs have further safety advantages and some disadvantages in comparison to solid-fuel reactors [17].

More modern lithium fluoride thorium reactor designs are being proposed as an alternative fuel cycle solution by several advocacy groups. A major problem with solid fuels is the accumulation of Xe-135, which acts as a short-lived neutron poison: in the MSR the Xe-135 is removed at the primary salt pump bowl. Introducing a small quantity of oxygen in the gas prevents corrosion of the piping [17]. TE USA is now working with Idaho National Laboratory to couple the IMSR to advanced industrial systems. Several systems have been designed and proposed. These can serve energy-intensive industries with stable heat and power for clean H_2, O_2 production, and by extension ammonia and methanol production.

Desalination is also a very significant market sector for IMSR heat and power. IMSR has the potential to be a transformative technology. When coupled with advanced industrial systems, IMSR enables new, transformative clean industries.

An IMSR power plant can rapidly load follow grid power demand and this power plant is anticipated to deliver power at less than $50 per MWh, which is highly competitive with fossil fuel combustion, thus will be a good candidate for electric power companies from their return on their investment point of view.

This type of high temperature reactor allows for much more than just electricity production. An IMSR power plant can deliver 600 °C heat by liquid salt up to 5 kilometers to an industrial energy park. This allows the IMSR baseload heat production of nuclear to be switched from electric power provision to the production of the most valuable high energy products in off-peak hours. This maximizes use of IMSR heat energy and allows the IMSR to run in the most capital efficient manner.

2.14.1 Thermal Storage and Desalination

Demands for safe, secure supplies of potable water globally are increasing faster than can be provided by natural, ever-depleting sources of fresh water. Simultaneously, global demand for electricity is also projected to grow significantly.

Desalination of seawater and brackish water is extremely energy intensive. The IMSR is uniquely suited to provide clean, heat energy and electric power on an industrial scale needed at cost-competitive prices to enable far greater deployment of desalination technologies today.

In addition to utilizing heat energy for desalination, hot industrial salts can be directed to a hot salt mass energy storage, a method that is already in use today. These hot salt thermal energy reservoirs supported by IMSR heat can be used as a grid sink for excess Wind and Solar electric power production. This system negates any need for grid-based electric power storage and is highly complementary to wind and solar power production. The cheap and effective salt-based thermal storage would act as an energy battery that will allow the demand curve to be supplied at the appropriate service levels without damaging surges taxing the grid system.

Studies conducted by TE USA and Idaho National Laboratory (INL) have shown that the IMSR power plants would be an effective system, relative to all other systems under review, to provide a growing water supply and stable power to the grid. The expanding growth demands on power and water can be served by an IMSR—a low-cost, carbon-free source of inherently safe energy.

2.14.2 H_2 from High Temperature Steam Electrolysis

Making H_2 from natural gas (NG) is the dominant method today but is highly sensitive to NG input prices. The (IMSR) is uniquely suited to provide a reliable and secure alternative method for H_2 production that has negligible input price volatility. The IMSR's can deliver the temperatures (600C+) and electric power that are needed for alternative methods for H_2 and O_2 production.

TE USA and INL have shown that the IMSR would be the most effective system of those reviewed to date to enable the best method of clean cost-competitive H_2 supply.

Analysis by INL and TE USA have shown that the IMSR is highly suited to be coupled to an industrial facility using high temperature steam electrolysis for H_2 production. Findings of the studies show that there are many other H_2, O_2, NH_3, and heat power production combinations that can be tailored to a great number of industrial applications.

2.14.3 Synthesized Transport Fuels

Production of transport fuels, including gasoline, using the IMSR, processes heat and electricity at a cost-competitive position with fossil fuels and represents a dramatic shift in economics of liquid fuel synthesis technology. This shift could have a profound effect on the industrial production methodologies of a broad range of valuable chemicals and fuels used in our industrial society. Demonstrating the production of synthetic gasoline at an industrial scale will certainly be followed closely by the production of other fuels such as aviation fuels, Liquefied Petroleum Gas (LPG), Diesel and others.

Note that: LPG is a portable, clean, and efficient energy source which is readily available to consumers around the world. LPG is primarily. The LPG should not be confused with Liquefied natural gas or associated petroleum gas. Liquefied Petroleum Gas (LPG) or Liquid Petroleum gas is a flammable mixture of hydrocarbon gases used as fuel in heating appliances, cooking equipment, and vehicles. [10]

Gasoline is also a symbolic fuel the public is familiar with and would give a clear signal of the immense opportunities that synthetically-derived fuels from nuclear power-driven process heat would represent. Namely: stabilized cost of energy inputs, sequestration of atmospheric carbon, and economic alternatives to fossil fuels. All of these opportunities are symbolic of the potential of IMSR to be a transformative technology and enables many new innovations and competitive clean industrial technologies that combine to drive economic growth and deep decarbonization of primary energy systems.

2.14.4 Ammonia Production Coupled to IMSR

During 2016, thirty plants produced 9.4 million metric tons of ammonia (NH_3), principally based on the Haber-Bosch reaction processes. The principal feedstock to these plants is natural gas, which is reformed with steam to produce a target stoichiometric gas mixture of CO_2, N_2, and H_2. Sorbents are used to remove CO_2 and other contaminants prior to synthesizing NH_3. Ammonia is used to produce a wide variety of fertilizers, nitric acid, fuels, and amine-based chemicals used broadly in industrial agriculture.

The above opportunities are examples of how IMSR can benefit the large and growing ammonia industry. Hydrogen that can be produced by high temperature steam electrolysis (HTSE) can replace the fossil-fuel intensive steam methane reforming technique. This would eliminate CO_2 emissions associated with hydrogen production today. The economics of HTSE when compared with fossil fuel-based hydrogen production, are based on the value of green house gas (GHG) emissions avoidance, as well as the market value of oxygen production for industrial uses, which represents a valuable byproduct of the HTSE process. Another possible opportunity for IMSR is for a modified interface with either a conventional or a revised steam methane reforming plant – a similar system to one used for methanol production. The significant benefits of a novel and disruptive NH_3 economy can be brought rapidly to fruition with the hybrid coupling of IMSR process heat with large scale ammonia production.

2.14.5 Coupling IMSR Technology into Direct Reduction Steel with H_2

It has been estimated, in the studies conducted at INL, that hydrogen-based high performance steel making could be cost-competitive with traditional steel production when coupled to an IMSR hybrid energy H_2 production system. This could also reduce total CO_2 emissions from steel production by 80 percent (Fischedick et al. 2014b) [11].

2.15 Conclusions

IMSR is a molten salt nuclear will enable affordable clean water, steel, hydrogen, and synthetic fuel [12].

IMSR nuclear island, produces 600 °C industrial heat, Balance-Of-Plant (BOP) can be a broad a board range of industrial applications—not just power provision.

Note that: Balance-of-plant is a term generally used in the context of power engineering to refer to all the supporting components and auxiliary systems of a power plant needed to deliver the energy, other than the generating unit itself. These may include transformers, inverters, supporting structures etc., depending on the type of plant. [13–15]

The IMSR nuclear island produces 600 °C industrial and balance-of-plant can be a broad range of industrial applications—not just power provision.

Finally, in conclusion the Canadian Company Terrestrial Energy Inc. (TEI), whose is building the Integral Molten Salt Reactor (IMSL) has secured $10M CDN in funding to develop an unusual type of nuclear reactor, based on Oakville, Ontario Canada. TEI will use the funds to develop this innovative technology in form an integrated system as branch of MSR, which as an old idea that was born at ORNAL yet has turned into IMSR innovative technical approach and with IMSR400 into production in a near future time frame.

The general layout of an IMSR plant. As outlined in the Stopping Climate Change page on the "Terrestrial Energy Integral Molten Salt Reactor (IMSR)," the design is a SMR design based closely on a DMSR design from the Oak Ridge National Laboratory, but also incorporates elements found in the SmAHTR, a later design from the same laboratory.

As outlined in the January 8th, 2016 Next Big Future post, "Terrestrial Energy gets funding for development for Game Changing Molten Salt Nuclear Reactor," IMSR reactors promise nuclear power that is far cheaper and greener than traditional methods.

IMSRs differ from traditional fission-based nuclear reactors in that they use fuel (in this case, denatured uranium) which has been dissolved in a molten liquid salt. Because an IMSR's fuel is in liquid form, it functions as both fuel and coolant, transporting heat away from the reactor as it circulates. Thus, an IMSR cannot go into meltdown because a loss of coolant (the traditional cause of meltdowns) would also mean a loss of the fuel needed to drive the reactor.

In addition to being *"melt down proof,"* IMSR technology offers other advantages:
- With the fuel and cooling systems essentially combined, IMSRs are a much simpler design than traditional fission reactors and so can be made smaller and cheaper.
- IMSRs use denatured uranium, a fuel that does not require extensive processing (i.e. lower fuel costs) and cannot be used to build nuclear weapons (no proliferation risks).
- IMSRs require only one-sixth the fuel needed annually by traditional reactors (lower operating costs).
- IMSRs can use recycled fuel.

For all of these advantages, IMSRs would still produce radioactive waste. However, this waste would be at far lower volume (kilograms versus tons) and be far shorter-lived (200–300 years vs millennia) when compared to traditional reactors.

References

[1] "IMSR starts second stage of Canadian design review - World Nuclear News". www.world-nuclear-news.org. (Accessed:17 October 2018).
[2] https://www.terrestrialenergy.com/.
[3] "Pre-Licensing Vendor Design Review - Canadian Nuclear Safety Commission". Nuclearsafety.gc.ca. (Accessed: 17 June 2018).
[4] "Terrestrial Energy to complete US loan guarantee application". world-nuclear-news.org. (Accessed: 12 December 2016).
[5] "Integrated Molten Salt Reactor passes pre-licensing milestone". world-nuclear-news.org. (Accessed: 30 January 2018).
[6] B. Zohuri, Compact Heat Exchangers: Selection, Application, Design and Evaluation, 1st Edition, Springer Publication Company, 2016.
[7] https://aris.iaea.org/PDF/IMSR400.pdf.
[8] J.R. Engel, W.W. Grimes, H.F. Bauman, H.E. McCoy, J.F. Bearing, W.A. Rhoades, "Conceptual design characteristics of a denatured molten salt reactor with once-through fueling" (PDF). ORNL-TM-7207.
[9] SmAHTR presentation by Sherrell Greene" (PDF). Archived from the original (PDF) on 2015-02-06. (Accessed: 6 February 2015).
[10] https://en.wikipedia.org/wiki/Thermal-neutron_reactor.
[11] M. Fischedick, J. Roy, A. Abdel-Aziz, A. Acquaye, J.M. Allwood, J.-P. Ceron, Y. Geng, H. Kheshgi, A. Lanza, D. Perczyk, L. Price, E. Santalla, C. Sheinbaum, K. Tanaka, Industry, in: O. Edenhofer, R. Pichs-Madruga, Y. Sokona, E. Farahani, S. Kadner, K. Seyboth, A. Adler, I. Baum, S. Brunner, P. Eickemeier, B. Kriemann, J. Savolainen, S. Schlömer, C. vonStechow, T. Zwickel, J.C. Minx (Eds.), Climate Change 2014: Mitigation of Climate Change. Contribution of Working Group III to the Fifth Assessment Report of the Intergovernmental Panel on Climate Change, Cambridge University Press, Cambridge, United Kingdom and New York, NY, USA, 2014.
[12] B. Zohuri, Hydrogen Energy: Challenges and Solutions for a Cleaner Future, 1st Edition, Springer Publishing Company, 2016.

[13] McGee, Chris. "Balance of Plant for Wind Projects". www.esru.strath.ac.uk. (Accessed: 29 June 2016). "Balance Of Plant" (PDF). Alstom. (Accessed: 29 June 2016).

[14] "NFCRC: FUEL CELL POWER PLANT: MAJOR SYSTEM COMPONENTS - Balance of Plant". www.nfcrc.uci.edu. Archived from the original on 11 October 2015. Accessed: (29 June 2016).

[15] R.C. Briant, et al., Nucl. Sci. Eng. 2 (1957) 797.

[16] P.N. Haubenreich, J.R. Engel, The Molten-Salt Reactor Experiment, Nucl. Appl. Tech. 8 (1970) 118.

[17] S. Peggs (BNL/ESS), W. Horak, T. Roser (BNL), G. Parks (Cambridge U.), M. Lindroos (ESS), R. Seviour (ESS/Huddersfield U.), S. Henderson (FNAL), R. Barlow, R. Cywinski (Huddersfield U.), J.-L. Biarrotte (IPN), A. Norlin (IThEO), V. Ashley, R. Ashworth (Jacobs), A. Hutton (JLab), H. Owen (Manchester U.), P. McIntyre (TAMU), J. Kelly (Thor Energy/WNA), Proceedings of IPAC2012, New Orleans, Louisiana, USA.

CHAPTER 3

New Approach to Energy Conversion Technology

A nuclear reactor produces and controls the release of energy from splitting the atoms of uranium. Uranium-fueled nuclear power is a clean and efficient way of boiling water to make steam which drives turbine generators. Except for the reactor itself, a nuclear power station works like most coal or gas-fired power stations.

3.1 Introduction

In the United States, most reactors design and development for the generation of electrical power was branched from early nuclear navy research, when it was realized that a compact nuclear power plant (NPP) would have a great advantages for submarine-driven nuclear propulsion system. To have such power plant on board would make possible long voyages across the oceans at high speeds without the necessity for resurfacing at frequent intervals.

Operating temperatures of conventional light water reactors (LWRs), 280–320 °C, limit power conversion systems (PCSs) to producing pressurized steam that drives a condensing steam turbine. After employing thermal recovery measures, nuclear plants using this Rankine cycle see a net plant efficiency of around 32–34%. Comparatively, gas turbines with turbine inlet temperatures of up to and greater than 1400 °C have simple cycle efficiencies of around 40% that can be boosted to around 60% in a combined cycle. The ability of the combined cycle or Brayton with recuperator cycle to drastically improve net plant efficiency is an especially appealing feature to employ with a nuclear power source, given the very low fuel costs for nuclear energy, but has previously been technically infeasible given the high operating temperature requirements of the combined cycle (Fig. 3.1).

One of the differentiating features of the fluoride salt high temperature reactors (HTRs) is the operating temperature range of the primary coolant loop, 600–700 °C (reactor inlet and outlet temperatures, respectively).

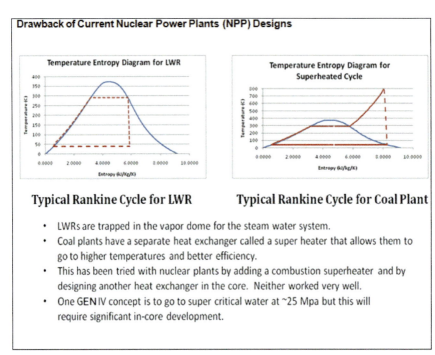

Fig. 3.1 Drawback of current nuclear power plants (NPP) designs [1].

Although other advanced, HTRs have been developed, the high temperature characteristics of the lithium fluoride and beryllium fluoride eutectic (flibe) molten salt primary coolant used in fluoride salt HTRs enables an operating temperature range that is uniquely suited to driving an open air combined cycle, which sees proportional increases in efficiency and power generation with elevated turbine inlet temperatures.

A study was conducted in September 2004 by a team of experts at the University of California, Nuclear Engineering Department [1]. The executive summary shows that the electrical PCS for the next-generation nuclear plant (NGNP) will take advantage of a significantly higher reactor outlet temperature to provide greater efficiency than can be achieved by the current generation of LWRs. In anticipation of the design, development, and procurement of an advanced PCS for NGNP, the study was initiated to identify the major design and technology options and their tradeoffs that must be considered in the evaluation of PCS options to support future research and procurement decisions. These PCS technology options affect cycle efficiency, capital cost, system reliability, and maintainability and technical risk, and therefore the cost of electricity (COE) from generation IV

(GEN IV) systems. A reliable evaluation and estimate of actual costs requires an optimized, integrated PCS design. At that early stage of the NGNP project it was useful to identify the technology options that would be considered in the design of proposed PCS systems, identify the system performance and cost implications of these design options, and provide a general framework for evaluating the design choices and technology tradeoffs.

The ultimate measure of the value of power-conversion options is the COE produced, which is a function of capital and operating cost recovery and the system efficiency and reliability. Evaluating cost is difficult to do without detailed integrated designs, but it is possible to identify the factors that influence component and system performance, cost, and technical risk. In this study, several existing Brayton conversion system designs were studied to illustrate and evaluate the implications of the major design choices to assess performance against the GEN IV economics and sustainability goals, and to identify areas of technical incompleteness or weakness. Several reference system designs were considered to provide a semiquantitative basis for performing comparisons. The reference systems included the gas turbine modular helium reactor (GT-MHR), pebble bed modular reactor (PBMR), GTHTR-300, Framatome indirect cycle design, and advanced high temperature reactor (AHTR) high temperature Brayton cycle designs. Where appropriate, Generations II, III, and III+ LWRs (two 1970s designs, the extended producer responsibility (EPR), and the economic simplified boiling water reactor (ESBWR)) were also considered.

The design choices and technology options considered relevant for the assessment of NGNP power-conversion options included the cycle types and operational conditions such as working fluid choices, direct versus indirect, system pressure, and interstage cooling and heating options. The cost and maintainability of the PCS is also influenced by the PCS layout and configuration including distributed versus integrated PCS designs, single versus multiple shafts, shaft orientation, and the implications for the pressure boundary.

From the summary in Table 3.1, it is apparent that high temperature gas reactor power-conversion design efforts to date have resulted in very different design choices based on project-specific requirements and performance or technical risk requirements.

In the review of existing designs and the evaluation of the major technology options, it immediately becomes apparent that the optimized design involves a complex tradeoff of diverse factors such as cost, efficiency, development time, maintainability, and technology growth path that must be considered in an integrated PCS system context before final evaluation.

Table 3.1 Summary of PCS design features for representative gas reactor systems.

Feature	PBMR (Horizontal)	GT-MHR	GTHTR300	Framatome Indirect	AHTR-IT
Thermal power (MWt)	400	600	600	600	2400
Direct vs indirect cycle	Direct	Direct	Direct	Indirect	Indirect
Recuperated vs combined cycle	Recuperated	Recuperated	Recuperated	Combined	Recuperated
Intercooled vs nonintercooled	Intercooled	Intercooled	Nonintercooled	Intercooled	Intercooled/reheat
Integrated vs distributed PCS	Distributed	Integrated	Distributed	Distributed	Distributed (modular)
Single vs multiple TaylorMade (TM) shafts	Single (previously Multiple)	Single	Single	Single	Multiple (modular)
Synchronous vs asynchronous	Reduction to synchronous	Asynchronous	Synchronous	Synchronous	Synchronous
Vertical vs horizontal TM	Horizontal	Vertical	Horizontal	Horizontal	Vertical
Submerged vs external generator	External	Submerged	Submerged	External	Submerged

General observations derived from the review of the reference systems, including comparisons with LWR systems where applicable, include following:
- There are key PCS design choices that can have large effects on PCS power density and nuclear island size, making careful and detailed analysis of design tradeoffs important in the comparison of PCS options.
- Considering the major construction inputs for nuclear plants—steel and concrete—HTRs appear to be able to break the economy-of-scale rules for LWRs and achieve similar material-input performance at much smaller unit sizes.
- For HTRs, a much larger fraction of total construction inputs goes into the nuclear island. To compete economically with LWRs, HTRs must find approaches to reduce the relative costs for nuclear-grade components and structures.

PCS technology options also include variations on the cycle operating conditions and the cycle type that can have an important impact on performance and cost. These options include following:
- Working fluid choice—He, N_2, CO_2, or combinations have been considered. Working-fluid physical characteristics influence cycle efficiency and component design.
- System pressure—Higher pressures lead to moderate efficiency increases and smaller PCS components but increase the pressure boundary cost—particularly for the reactor vessel—which introduces a component design, and a system cost and performance tradeoff.
- Direct versus indirect—Indirect cycles involve an intermediate heat exchanger (IHX) and resulting efficiency reduction, and more complex control requirements, but facilitate maintenance.
- Interstage cooling (or heating) results in higher efficiency but greater complexity.

Some of the observations from this assessment of these factors include following:
- Differences between He versus N_2 working fluids were not considered critical for turbomachinery design, because both involve similar differences from current combustion turbines, with the primary difference being in the heat exchanger (HX) size to compensate for the lower N_2 thermal conductivity.
- N_2 allows 3600-rpm compressor operation at thermal powers at and below 600 MW(t), while He compressors must operate at higher speeds requiring reduction gears, asynchronous generators, or multiple-shaft

configurations. However, power up-rating to approximately 800 MW(t) would permit 3600-rpm He compressor operation, providing a potentially attractive commercialization approach. Turbomachinery tolerances for He systems do not appear to be a key issue.
- Direct /Indirect—Efficiency loss can be 2–4 %, depending on design, and the IHX becomes a critical component at high temperatures. Direct cycles have an extended nuclear grade pressure boundary. Maintainability is considered a key design issue for direct cycles.
- Interstage cooling, as well as bottoming cycles (Rankine), can result in significant efficiency improvements, but at a cost of complexity and lower temperature differences for heat rejection, affecting the potential for dry cooling and reduced environmental impact from heat rejection.

The PCS configuration and physical arrangement of the system components has important effects on the volume and material inputs into structures, on the pressure boundary volume and mass, gas inventories and storage volume, the uniformity of flow to HXs, pressure losses, and maintainability. The major factors considered in this study included following:
- Distributed versus integrated PCS design approach—PCS components can be located inside a single pressure vessel (e.g., GT-MHR), or can be divided between multiple pressure vessels (e.g., PBMR, HTR-300). This is a major design choice, with important impacts in several areas of design and performance.
- Shaft orientation (vertical/horizontal)—Orientation affects the compactness of the system, the optimal design of ducting between turbomachinery and HXs. Vertical turbomachinery provides a reduced PCS footprint area and building volume and can simplify the ducting arrangement to modular recuperator and intercooler HXs.
- Single versus multiple shafts—Single shafts may include flexible couplings or reduction gears. In multiple-shaft systems, turbo-compressors are separated from synchronous turbo-generators, allowing the compressors to operate at a higher speed and reducing the number of compressor stages required. Multiple shafts and flexible couplings reduce the weight of the individual turbomachines that bearings must support.
- Pressure boundary design—The pressure vessels that contain the PCS typically have the largest mass of any PCS components and provide a significant (~33%) contribution to the total PCS cost.

3.2 Waste Heat Recovery

Waste heat is heat generated in a process by way of fuel combustion or chemical reaction, which is then "dumped" into the environment and not reused for useful and economic purposes. The essential fact is not the amount of heat, but rather its "value." The mechanism to recover the unused heat depends on the temperature of the waste heat gases and the economics involved.

Large quantities of hot flue gases are generated from boilers, kilns, ovens, and furnaces. If some of the waste heat could be recovered then a considerable amount of primary fuel could be saved. The energy lost in waste gases cannot be fully recovered. However, much of the heat could be recovered and adopting the following measures as outlined in this chapter can minimize losses.

3.3 PCS Components

The effectiveness or efficiency of the major PCS components, primarily the HXs and turbomachinery, is clearly a major factor in system cost and performance. Overall, summary of each of these components is given in the following sections.

3.3.1 Heat Exchangers

An HX is a device built for efficient heat transfer from one medium to another, whether the media are separated by a solid wall so that they never mix, or the media are in direct contact. They are widely used in space heating, refrigeration, air conditioning, power plants, chemical plants, petrochemical plants, petroleum refineries, and natural gas processing. One common example of an HX is the radiator in a car, in which a hot engine-cooling fluid, like antifreeze, transfers heat to air flowing through the radiator.

For efficiency, HXs are designed to maximize the surface area of the wall between the two fluids, while minimizing resistance to fluid flow through the exchanger. The exchanger's performance can also be affected by the addition of fins or corrugations in one or both directions, which increase surface area and may channel fluid flow or induce turbulence.

A we mentioned the HX are transferring heat from one fluid to another and in particular at compact size is one of the important aspects of PCS components, if we want to rely on combined cycles–driven efficiency of the next-generation NPPs.

High heat transfer rates, modular construction, exotic alloys, low hold-up volume, and high internal turbulence, all combined to improve efficiencies, and process control, with longer maintenance cycles. The temperature difference at the pinch depends on the decision of HX; in general, the smaller the temperature difference the more expensive the HX [2].

Taking Fig. 3.2 under consideration, a pinch point is defined as the point where the temperature difference is a minimum. Taking as an example a temperature difference of 20 K, say, then the combined cold stream can be moved from left to right on the diagram until the temperature difference at the pinch point is 20 K as shown in Fig. 3.2. It can then be seen that external heating of 90 kW and external cooling of 140 kW are required; all other energy changes can be achieved by HXs between the various streams; the difference between the external heating and the external cooling is (90 - 140) = -50 kW, the same as in Table 3.1. Note that the process in a boiler or condenser would appear as a horizontal line on a diagram such as Fig. 3.2. The data are given in Table 3.2.

As we said compact HXs have a vital role to produce higher efficiency if they are used in a combined gas-steam power cycles versus conventional shell-and-tube HXs.

They are available with standard connections in various diameters and in different lengths. Compact HXs can be designed in horizontal and

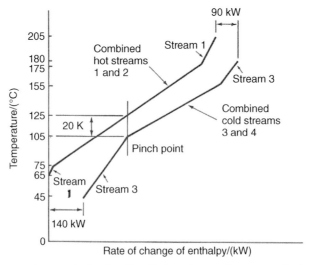

Fig. 3.2 Temperature against rate of enthalpy change for composite hot and cold streams.

Table 3.2 Data for four fluid streams.

Stream number	Initial temp. (°C)	Final temp. (°C)	Mass flow rate (kg/s)	Specific heat cap. (kJ/ kg K)	Heat cap. rate (kW/ K)	Rate of enthalpy increase (kW)
1	205	65	2.00	1.00	2.0	−280
2	175	75	3.20	1.25	4.0	−400
3	45	180	3.75	0.80	3.0	+405
4	105	155	3.00	1.50	4.5	+225 −50

vertical orientation and can be used for a wide variety of applications including, among others and they are listed below:
- Compressed gas/water coolers
- Water/water coolers
- Oil/water coolers
- Preheaters
- Condensers and evaporators for chemical and technical processes of all kinds.
- Machine coolers
- Oil coolers for hydraulic systems
- Oil and water coolers for power machines
- Refrigeration and air-conditioning units.

If we look at the HX application in case of combined gas–steam power cycles (Fig. 3.3), where objective of this cycle is to combine a Brayton cycle with a Rankine cycle. This combination improves significantly the thermal efficiency of the power plant.

The idea behind a combined gas–steam power cycle is to use the exhaust high temperature gases from the gas turbine to heat the steam within the boiler of the steam turbine.

If we consider such combined cycle, the T-s diagram is presented by Fig. 3.4.

HX components are defined, and the required designs are summarized as follows:
- HXs designs have significant impacts on both the efficiency and cost of the PCS. For a given HX type, higher effectiveness must be balanced against the increased size or pressure drop implications. Using small passages increases HX surface area per unit volume, but those same small

94 Molten salt reactors and integrated molten salt reactors

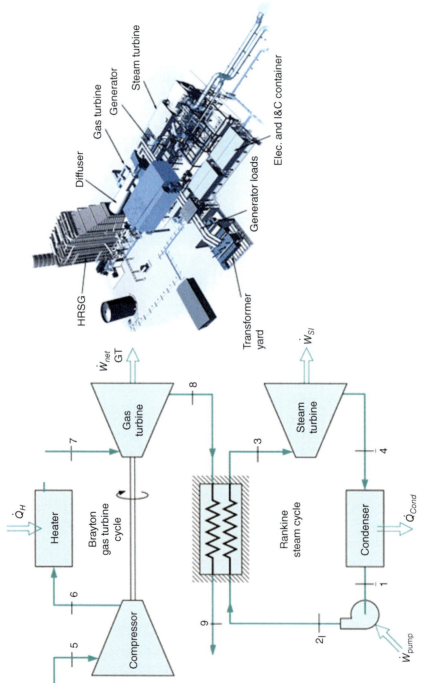

Fig. 3.3 Combined gas (Brayton)–steam (Rankine) cycle.

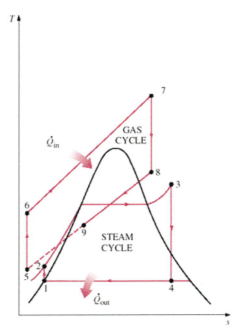

Fig. 3.4 *T-s* diagram of a combined gas (Brayton)– steam (Rankine) cycle.

passages tend to reduce the heat transfer coefficient due to laminar flow. Higher pressures may be utilized to force those flows back into the turbulent region, but those higher pressures force construction of a more robust pressure boundary and increase pumping power.

- The recuperator effectiveness and total HX pressure drop is a significant impact on the cycle efficiency and there is significant leverage in optimizing the recuperator design for both high heat transfer effectiveness and minimum pressure drop. For modular recuperators, careful attention must be paid to the module configuration and duct design to obtain equal flow rates to each module.
- Material limitations may limit the operating temperatures for many components, including the reactor vessel and HXs. But because of the large flexibility of the Brayton cycle, high-efficiency systems can still be designed within these limitations. Fabrication techniques will probably differ between intermediate, pre-, and intercooler, and recuperator HXs, because of their operating temperature ranges. It would appear that transients could be tolerated by most of these HX designs.

An HX is a device for transferring heat from one fluid to another. There are three main categories:
1. *Recuperative*: in which the two fluids are at all times separated by a solid wall;
2. *Regenerative*: in which each fluid transfers heat to or from a matrix of material;
3. *Evaporator*: in which the enthalpy of vaporization of the fluid is used to provide a cooling effect.

The most commonly used type is the recuperative HX and they can be classified by flow arrangement, where numerous possibilities exist for flow arrangement in HXs.

In this type of HXs, the two fluids can flow [3–7]
1. in *parallel flow*,
2. in *counter flow*, or
3. in *cross flow*.

HXs may be classified according to their flow arrangement.
1. *Parallel flow*: In parallel flow HX configuration, the hot and cold fluids enter at the same end of the HX, flow through in the same direction (i.e., in parallel to one another to the other side), and leave together at the other end, as illustrated in Fig. 3.5A [4].

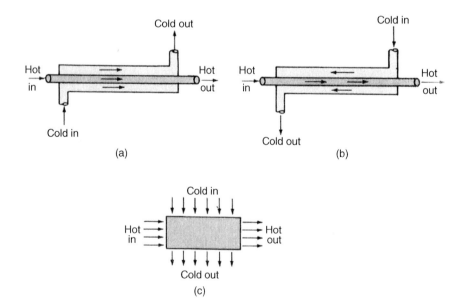

Fig. 3.5 (A) Parallel flow, (B) counter flow, and (C) cross flow [3].

2. *Counter flow:* In counter-flow HX configuration, the hot and cold fluids enter in the opposite ends of the HX and flow through in opposite directions, as illustrated Fig. 3.5B. [5]
3. *Cross flow:* In the cross-flow HX configuration, the two fluids usually flow at right angles to each other, as illustrated in Fig. 3.5C [4]. In the cross-flow arrangement, the flow may be called *mixed* or *unmixed*, depending on the design. The counter current design is most efficient, in that it can transfer the most heat. In a cross-flow HX, the fluids travel roughly perpendicular to one another through the exchanger.

Fig. 3.6A shows a simple arrangement in which both hot and cold fluids flow through individual channels formed by corrugation; therefore, the fluids are not free to move in the transverse direction. Then each fluid stream is said to be unmixed. Fig. 3.6B illustrates a typical temperature profile for the outlet temperatures when both fluids are unmixed, as shown in Fig. 3.6A.

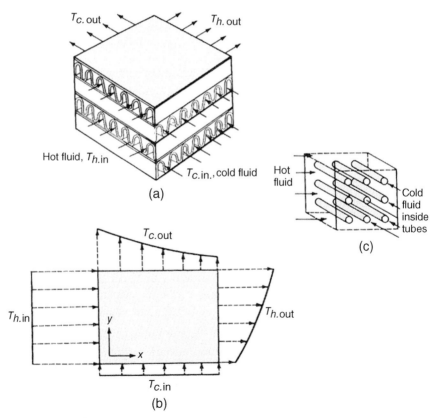

Fig. 3.6 Cross-flow arrangements: (A) Both fluids unmixed; (B) temperature profile when both fluids are unmixed; (C) cold fluid unmixed, hot fluid mixed [3].

The inlet temperatures for both fluids are assumed to be uniform, but the outlet temperatures exhibit variation transverse to the flow.

In the flow, arrangement shown in Fig. 3.6C, the cold fluid flows inside the tubes and so is not free to move in the transverse direction. Therefore, the cold fluid is said to be unmixed. However, the hot fluid flows over the tubes and is free to move in the transverse direction. Therefore, the hot fluid stream is said to be mixed. The mixing tends to make the fluid temperature uniform in the transverse direction; therefore, the exit temperature of a mixed stream exhibits negligible variation in the crosswise direction.

In general, in a cross-flow exchanger, three idealized flow arrangements are possible:
1. Both fluids are unmixed.
2. One fluid is mixed, and the other is unmixed.
3. Both fluids are mixed.

The last arrangement is not commonly used.

In a shell-and-tube exchanger, the presence of a large number of baffles serves to "mix" the shell-side fluid in a sense discussed above; that is, its temperature tends to be uniform at any cross-section.

The driving temperature across the heat transfer surface varies with position, but an appropriate means temperature can be defined. In most simple systems this is the logarithmic mean temperature difference (LMTD). Sometimes direct knowledge of the LMTD is not available and the number of transfer units (NTU) method is used.

In summary, the NTU method is used to calculate the rate of heat transfer in HXs (especially counter current exchangers) when there is insufficient information to calculate the LMTD. In HX analysis, if the fluid inlet and outlet temperatures are specified or can be determined by a simple energy balance, the LMTD method can be used; but when these temperatures are not available The NTU or the *effectiveness method* is used.

Now, we summarize here the principle ones as shown in Fig. 3.5A–C. The temperature variations of each type are shown in Fig. 3.5A–C. It can also be shown that for all three cases of Fig. 3.5 the true temperature difference is the *LMTD* given by [4–6]:

$$LMTD = \frac{(\Delta t_1 - \Delta t_2)}{\ln\left(\Delta t_1 / \Delta t_2\right)} \quad (3.1)$$

On the other hand, in case of NTU or effectiveness method analysis, to define the effectiveness of an HX we need to find the maximum possible heat transfer that can be hypothetically achieved in a counter-flow HX of an infinite length. Therefore, one fluid will experience the maximum possible temperature difference, which is the difference of $T_{hot,inlet} - T_{cold,inlet}$ (The temperature difference between the inlet temperature of the hot stream $T_{hot,inlet} = T_{h,i}$ and the inlet temperature of the cold stream $T_{cold,inlet} = T_{c,i}$). The method proceeds by calculating the heat capacity rates (i.e., mass flow rate multiplied by specific heat) C_h and C_c for the hot and cold fluids, respectively and denoting the smaller one as C_{min}. Studies of Ozisik [4], Incropera and DeWitt [5], and Incropera et al. [6] presented a detailed analysis of the NTU method and we just write the result of finding the effectiveness that is given as

$$E = 1 - \exp[-NTU] \qquad (3.2)$$

As we said, for more detailed analysis of these two methods readers should refer to the above-mentioned classical

3.3.1.1 Recuperative HXs

Industrial process facilities recover, or "recuperate" otherwise wasted heat energy using HXs, but on a much larger scale.

A recuperator is a special-purpose counter-flow energy recovery HX positioned within the supply and exhaust air streams of an air-handling system, or in the exhaust gases of an industrial process, to recover the waste heat.

In many types of processes, combustion is used to generate heat, and the recuperator serves to recuperate, or reclaim this heat, to reuse or recycle it. The term recuperator refers as well to liquid-liquid counter-flow HXs used for heat recovery in the chemical and refinery industries and in closed processes such as ammoni5-water or LiBr-water absorption refrigeration cycles.

Recuperators are often used in association with the burner portion of a heat engine to increase the overall efficiency. For example, in a gas turbine engine, air is compressed, mixed with fuel, which is then burned and used to drive a turbine. The recuperator transfers some of the waste heat in the exhaust to the compressed air, thus preheating it before entering the fuel burner stage. As the gases have been preheated, less fuel is needed to heat the gases up to the turbine inlet temperature. By recovering some of the energy usually lost as waste heat, the recuperator can make a heat engine or gas turbine significantly more efficient.

In a recuperator, heat exchange takes place between the flue gases and the air through metallic or ceramic walls. Ducts or tubes carry the air for combustion to be preheated; the other side contains the waste heat stream. A recuperator for recovering waste heat from flue gases is shown in Fig. 3.7.

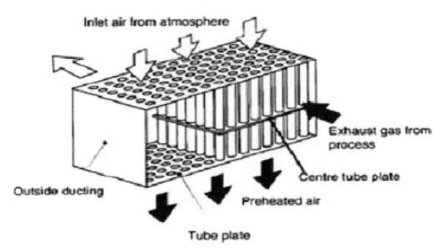

Fig. 3.7 Waste heat recovery using recuperator.

There are different type of manufactured recuperator depending on their application, and they are described in the following sections.

Fig. 3.8 is illustration of various recuperators manufactured by industries.

3.3.1.1.1 Metallic Radiation Recuperator

The simplest configuration for a recuperator is the metallic radiation recuperator, which consists of two concentric lengths of metal tubing as shown in Fig. 3.9.

The inner tube carries the hot exhaust gases while the external annulus carries the combustion air from the atmosphere to the air inlets of the furnace burners. The hot gases are cooled by the incoming combustion air, which now carries additional energy into the combustion chamber. This is the energy, which does not have to be supplied by the fuel; consequently,

Vertical Recuperator

Recuperator Inside

Horizontal Recuperator

Fig. 3.8 Illustration of different types of recuperator.

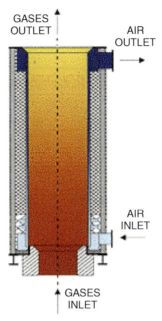

Fig. 3.9 Metallic radiation recuperator.

less fuel is burned for a given furnace loading. The saving in fuel also means a decrease in combustion air and therefore, stack losses are decreased not only by lowering the stack gas temperatures but also by discharging smaller quantities of exhaust gas.

The radiation recuperator gets its name from the fact that a substantial portion of the heat transfer from the hot gases to the surface of the inner tube takes place by radiative heat transfer. The cold air in the annuals, however, is almost transparent to infrared radiation so that only convection heat transfer takes place to the incoming air. As shown in the diagram, the two gas flows are usually parallel, although the configuration would be simpler, and the heat transfer would be more efficient if the flows were opposed in direction (or counter flow). The reason for the use of parallel flow is that recuperators frequently serve the additional function of cooling the duct carrying away the exhaust gases and consequently extending its service life.

3.3.1.1.2 Convective Recuperator

A second common configuration for recuperator is called the tube type or convective recuperator. As seen in the Fig. 3.10, the hot gases are carried through a number of parallel small-diameter tubes, while the incoming air

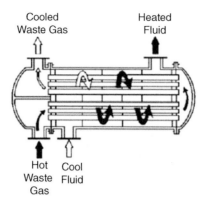

Fig. 3.10 Convective recuperator.

to be heated enters a shell surrounding the tubes and passes over the hot tubes one or more times in the direction normal to their axes.

If the tubes are baffled to allow the gas to pass over them twice, the HX is termed a two-pass recuperator; if two baffles are used, a three-pass recuperator, etc.

Although baffling increases both the cost of the exchanger and the pressure drop in the combustion air path, it increases the effectiveness of heat exchange. Shell- and tube-type recuperators are generally more compact and have a higher effectiveness than radiation recuperators, because of the larger heat transfer area made possible through the use of multiple tubes and multiple passes of the gases.

Using steel plain tubes, in these recuperators the heat transfer between the primary and secondary fluid is made by means of convection (Fig. 3.11).

3.3.1.1.3 Hybrid Recuperator

For maximum effectiveness of heat transfer, hybrid recuperators are used. These are combinations of radiation and convective designs, with a high-temperature radiation section followed by a convective section (See Fig. 3.12).

These are more expensive than simple metallic radiation recuperators but are less bulky.

3.3.1.1.4 Ceramic Recuperator

The principal limitation on the heat recovery of metal recuperators is the reduced life of the liner at inlet temperatures exceeding 1100 °C. To

New approach to energy conversion technology 103

Fig. 3.11 Typical commercial cross-flow convective recuperators.

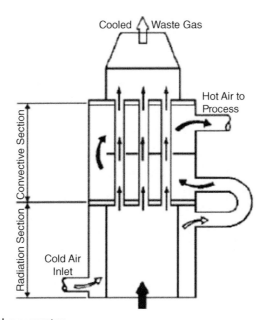

Fig. 3.12 Hybrid recuperator.

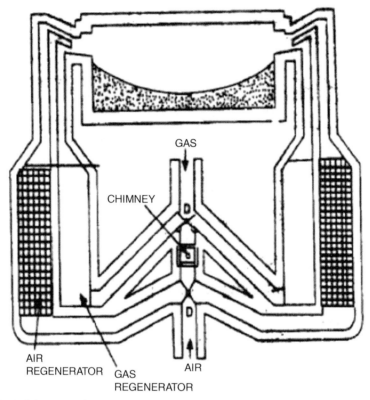

Fig. 3.13 Schematic of ceramic recuperator.

overcome the temperature limitations of metal recuperators, ceramic tube recuperators have been developed whose materials allow operation on the gas side to be at 1550 °C and on the preheated air side to be 815 °C on a more or less practical basis. Early ceramic recuperators were built of tile and joined with furnace cement, and thermal cycling caused cracking of joints and rapid deterioration of the tubes. Later developments introduced various kinds of short silicon carbide tubes, which can be joined by flexible seals located in the air headers (Fig. 3.13).

Earlier designs had experienced leakage rates from 8% to 60%. The new designs are reported to last 2 years with air preheat temperatures as high as 700 °C, with much lower leakage rates.

3.3.1.2 *Regenerative HXs*
A regenerative HX, or more commonly a regenerator, is a type of HX where heat from the hot fluid is intermittently stored in a thermal storage

medium before it is transferred to the cold fluid. To accomplish this, the hot fluid is brought into contact with the heat storage medium, then; the fluid is displaced with the cold fluid, which absorbs the heat.

In regenerative HXs, the fluid on either side of the HX can be the same fluid. The fluid may go through an external processing step, and then it is flowed back through the HX in the opposite direction for further processing. Usually the application will use this process cyclically or repetitively.

The regenerator represents class of HXs in which heat is alternately stored and removed from a surface. This heat transfer surface is usually referred as the matrix of the regenerator and for continuous operation; the matrix must be moved into and out of the fixed hot and cold fluid streams. In this case, the regenerator is called a rotary regenerator. If, on the other hand, the hot and cold fluid streams are switched into and out of the matrix, the regenerator is referred as a fixed matrix regenerator.

Fig. 3.14 is illustration of typical regenerator HXs and as it can be seen that Fig. 3.14A is depiction of rotary, Fig. 3.14B fixed-matrix, and finally Fig. 3.14C shows a rotating hoods of this type of HX.

A typical rotary regenerators or heat wheels are demonstrated in Fig. 3.15.

Any further detailed analysis, including advantages and disadvantages of this type of HX, is beyond the intended scope of this book and we refer readers to do their own research and investigations.

3.3.1.3 Evaporative HXs

Normally this type of HX can be found in any cooling tower of power plants of all types and as the name of this HX sounds, it uses evaporation process to the heat transfer processes.

In an open circuit cooling tower, warm water from the heat source is evenly distributed via a gravity or pressurized nozzle system directly over a heat transfer surface called "fill" or "wet deck," while air is simultaneously forced or drawn through the tower, causing a small percentage of the water to evaporate. The evaporation process removes heat and cools the remaining water, which is collected in the tower's cold water basin and returned to the heat source (typically a water-cooled condenser or other HX). Fig. 3.16 is illustration of a cooling tower along with evaporative HX.

Maintenance frequency will depend largely upon the condition of the circulating water, the cleanliness of the ambient air used by the tower, and the environment in which the tower is operating.

106 Molten salt reactors and integrated molten salt reactors

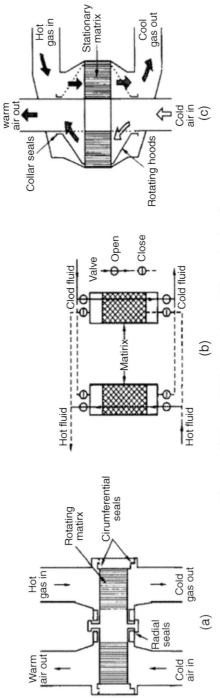

Fig. 3.14 (A) Rotary, (B) fixed matrix, and (C) rotating hoods.

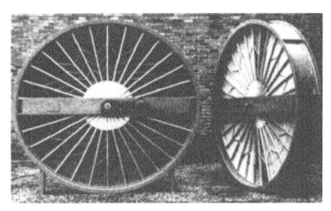

Fig. 3.15 Typical rotary regenerators or heat wheels.

3.3.2 Compact HXs

A type of HX that is specifically designed to realize the larger heat transfer area per unit volume is designated or known as compact HX. Compact HXs are commonly used in gas-to-gas and gas-to-liquid (or liquid-to-gas) HXs to counteract the low heat transfer coefficient associated with gas flow with increased surface area.

Fig. 3.17 is illustration of typical configuration and structure of a gas-to-liquid compact HX for a residential air conditioning system.

Compact HXs are a class of HXs that incorporate a large amount of heat transfer surface area per unit volume, where the area density β parameter comes to play.

The ratio of the heat transfer surface area of an HX to its volume is called the area density β. An HX with $\beta = 700$ m^2/m^3 (or 200 ft^2/ft^3) is classified as being compact. Examples of compact HXs are car radiators (1000 m^2/m^3), glass ceramic gas turbine HXs (6000 m^2/m^3), the regenerator of a Stirling engine (15,000 m^2/m^3), and the human lung (20,000 m^2/m^3).

An HX with $\beta > 700$ m^2/m^3 is classified as being compact HX, for example
- Car radiator ($\beta \approx 1000$ m^2/m^3)
- Gas ceramic gas turbine HX ($\beta \sim 6000$ m^2/m^3)
- Human lung ($\beta \approx 20,000$ m^2/m^3)

Most automotive HXs would come into the compact HX category as space is an extreme constraint for automotive applications. By the same talking recent research and study of open air combined cycles at University of New Mexico by Zohuri and McDaniel, et al. [8–11] and CO_2 closed

108   Molten salt reactors and integrated molten salt reactors

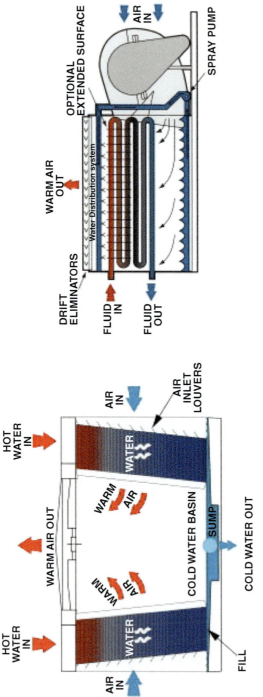

Fig. 3.16 Illustration of a cooling tower along with evaporative heat exchanger.

Fig. 3.17 A residential gas-to-liquid compact heat exchangers.

combined cycle by a team of engineers at Sandia National Laboratory [12–14] and other folks across universities and industry, utilizing the Brayton and Rankine cycles as a means of driving the efficiencies on the next-generation (GEN IV) NPP with a small modular reactors (SMR) approaches, show a promising application of this types of HXs.

Types of compact HXs are categorized as follows [14,15]:
- Tubular HX
- Fin-plate HX
- Tube-fin HX
- Plate-frame HX
- Regenerative HX

Designing this type of HXs is an easy task and the reasoning to argue why this not that simple, can be summarized below:
- Perform the required heat transfer and
  - Minimize size and weight
  - Minimize pressure drop
  - Meet required life
  - Be resistant to fouling and contamination
  - Minimize cost.

Bear in your mind, another important factor for designing these types of HXs (compact) in particular in their application for high-temperature combined cycle for the next-generation NPPs is pinch point and imposed golden rules driven by pinch technology, which will be mentioned in Chapter 5, which makes them very expensive to manufacture.

Comparing the compact HXs with traditional shell and tube HXs can be summarized below:
- They occupy 80% less space as compared shell and tube HX.
- Overall heat transfer coefficient is three or four times higher compared to shell and tube HX. This is very important advantage that is required by combined cycles in SMR to drive to higher thermal efficiency out from cost of ownership to make them more desirable.
- Low pressure and low temperature devices.

Limitations of the compact HXs can be worth mentioning are as follows:
- Maximum design pressure: 25 Bar.
- Maximum design temperature: 200 °C.
- It is only used for clean fluids.

For further detailed information and related tables as well as correlated experimental data we encourage readers to refer to classical and well-known book by Kays and London [15] as well as book edited by Shah [16].

3.4 Development of Gas Turbine

The gas turbine has experienced phenomenal progress and growth since its first successful development in the 1930s. The early gas turbines built in the 1940s and even 1950s had simple-cycle efficiencies of about 17% because of the low compressor and turbine efficiencies and low turbine inlet temperatures due to metallurgical limitations of those times. Therefore, gas turbines found only limited use despite their versatility and their ability to burn a variety of fuels. The efforts to improve the cycle efficiency concentrated in three areas are as follows:
1. Increasing the turbine inlet (or firing) temperatures.
2. Increasing the efficiencies of turbomachinery components.
3. Add modifications to the basic cycle. The simple-cycle efficiencies of early gas turbines were practically doubled by incorporating intercooling, regeneration (or recuperation), and reheating. The back work ratio of a gas-turbine cycle improves as a result of intercooling and reheating. However, this does not mean that the thermal efficiency will also improve. Intercooling and reheating will always decrease the thermal efficiency unless they are accompanied by regeneration. This is because intercooling decreases the average temperature at which heat is added, and reheating increases the average temperature at which heat is rejected. Therefore, in gas-turbine power plants, intercooling and reheating

are always used in conjunction with regeneration. These improvements, of course, come at the expense of increased initial and operation costs, and they cannot be justified unless the decrease in fuel costs offsets the increase in other costs. In the past, the relatively low fuel prices, the general desire in the industry to minimize installation costs, and the tremendous increase in the simple-cycle efficiency due to the first two figures [Fig. 3.19 T-s diagram] increased efficiency options to approximately 40% left little desire for incorporating these modifications. With continued expected rise in demand and cost of producing electricity, these options will play an important role in the future of gas-turbine power plants. The purpose of this book is to explore this third option of increasing cycle efficiency via intercooling, regeneration, and reheating.

Gas turbines installed until the mid-1970s suffered from low efficiency and poor reliability. In the past, large coal and NPPs dominated the base-load electric power generation [Point 1 in Fig. 3.18]. Base-load units are online at full capacity or near-full capacity almost all of the time. They are not easily or quickly adjusted for varying large amounts of load because of their characteristics of operation [16]. However, there has been a historic shift toward natural gas–fired turbines because of their higher efficiencies, lower capital costs, shorter installation times, better emission characteristics, the abundance of natural gas supplies, and shorter start up times [Point 1 in Fig. 3.18]. Now electric utilities are using gas turbines for base-load power production as well as for peaking, making capacity at maximum

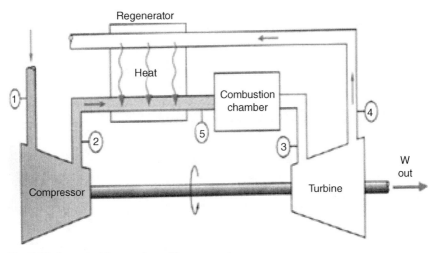

Fig. 3.18 A gas-turbine engine with recuperator.

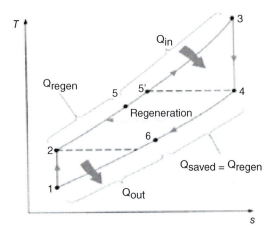

Fig. 3.19 T-s diagram of a Brayton cycle with regeneration.

load periods and in emergency situations because they are easily brought online or offline [2 in Fig. 3.18]. The construction costs for gas-turbine power plants are roughly half that of comparable conventional fossil fuel steam power plants, which were the primary base-load power plants until the early 1980s, but peaking units are much higher in energy output costs. A recent gas turbine manufactured by General Electric uses a turbine inlet temperature of 1425 °C (2600 °F) and produces up to 282 MW while achieving a thermal efficiency of 39.5% in the simple-cycle mode. Over half of all power plants to be installed in the foreseeable future are forecast to be gas turbine or combined gas-steam turbine types [Fig. 3.18].

Overall combined gas-steam power cycles were touched upon Section 6.3.1 and Fig. 3.20 is demonstration of difference between a gas and steam turbines blades.

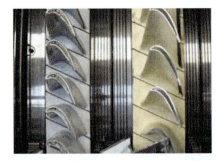 

Fig. 3.20 Difference between gas turbine (left) and steam turbine (right) blades.

Typical numerical examples for Brayton, Rankine, and combined cycles are presented below and they are supported by a computer code that was developed by Zohuri and McDanie [2] at University of New Mexico, Nuclear Engineering Department. The flow chart and overall capability of this code based on a static approach is explained in Chapters 8–10.

Numerical example:
- Simple Brayton efficiency: 36.8%
- Simple Rankine efficiency: 35%
- Combined Brayton and Rankine cycles: 56.4%

3.5 Turbomachinery

We have used thermal engines widely since they invented in the seventeenth century. There are many kinds of the engine, and they are used in our life. The two turbomachines that we face in nuclear plants are:
- gas turbine
- steam turbine.

Our main focus will be on gas turbine where Brayton cycle plays a big role and we extensively touched upon it in Zohuri and McDaniel [2] (Chapter 7) and it is worth to briefly talk about steam turbine here.

Fig. 3.21 is depiction of steam turbine layout in a simple form. The steam turbine has rotating blades instead of the piston and the cylinder of the reciprocating steam engine. This engine is used as the power source in the thermal and NPPs. The steam turbine utilizes dynamic pressure of the steam, and converts a thermal energy to a mechanical energy, though the reciprocating steam engine utilizes the static pressure of the steam. The both engines use the energy that is obtained at the expansion of the steam.

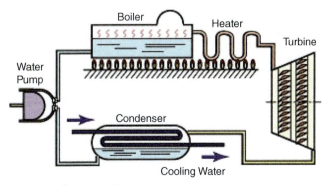

Fig. 3.21 Schematic of steam turbine.

Turbomachinery used in the new generation of nuclear power systems (GEN-IV) plays a significant role in commercial applications to produce the electricity of the future.

It is also good to know that any time we talk about steam, we will be dealing with Rankine cycle as a rule of thumb.

- First-order estimates of key turbine and compressor design and performance characteristics can be made with low-level analysis. For the reference systems, key turbomachinery design parameters, (speed, stages, stage diameters, blade heights, blade clearances) will be similar to current commercial gas turbine engines.
- At lower reactor thermal powers, He compressors will require greater than 3600 rpm operation to achieve efficiency goals (800 MW(t) allows 3600 rpm operation).
- Maximum system temperatures in the reference designs are near the limit for uncooled turbines.
- For both direct and indirect designs, the seals, housing, and bearing components will be fundamentally different than current gas turbines, requiring extensive development with the associated cost and risk.

These observations illustrate the complex interactions of the many design choices that will be considered in the NGNP PCS. It is clear that detailed and integrated design efforts must be performed on candidate designs before quantitative evaluations are possible. The assessment described in that study helped illuminate those critical design choices and the resulting implications for the cost and performance of the future NGNP PCS design.

3.6 Heat Transfer Analysis

Further analysis of heat transfer for any recuperative HX shows that the following governing equation stands by

$$Q = \dot{m}_H c_H (t_{H1} - t_{H2}) = \dot{m}_c c_c (t_{C1} - t_{C2}) = UA_o(LMTD)K \quad (3.3)$$

where $\dot{m}_H$ and $\dot{m}_C$ are the mass flow rates of the hot and cold fluids; c_H and c_C are the specific heats of the hot and cold fluids; A_0 is the outside area of the wall separating the two fluid; U is the overall heat transfer coefficient based on the outside area; K is a multiplying factor for cross-flow and mixed-flow types ($K=1$ for counter flow and parallel flow).

The temperature variations as shown in Fig. 3.22A–C can be shown for all three cases, where Fig. 3.22A shows temperature variations in counter

New approach to energy conversion technology 115

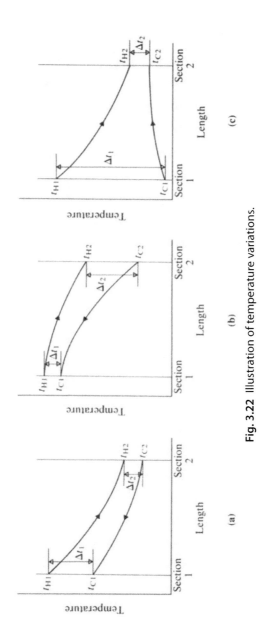

Fig. 3.22 Illustration of temperature variations.

flow with $\dot{m}_C c_C > \dot{m}_H c_H$ and Fig. 3.22B indicates temperature variations in counter flow with $\dot{m}_C c_C < \dot{m}_H c_H$ while Fig. 3.22C illustrates the temperature variations in parallel flow.

Using the concept of thermal resistance in series we have:

$$\frac{1}{UA_o} = \frac{1}{h_i A_i} + R_w + \frac{1}{h_o A_o} + F_i + F_o \qquad (3.4)$$

where subscription o and i refer to the outside and inside surfaces of the separating wall; where F represents fouling factor that for most HXs is a deposit of salts, oil, or other contaminant that will gradually build up on the heat transfer surfaces. This is allowed for in the design by using a *fouling factor*, F, in the form of an additional thermal resistance as in Eq. (3.3). Cleaning of the HX takes place when the fouling has reached the design value.

3.7 Combined-Cycle Gas Power Plant

A combined-cycle gas turbine (CCGT) power plant is essentially an electrical power plant in which a gas turbine and a steam turbine are used in combination to achieve greater efficiency than would be possible independently. The gas turbine drives an electrical generator while the gas turbine exhaust is used to produce steam in an HX, called a heat recovery steam generator (HRSG), to supply a steam turbine whose output provides the means to generate more electricity. If the steam were used for heat then the plant would be referred to as a cogeneration plant.

It is important first to distinguish between a closed cycle power plant (or heat engine) and an open cycle power plant. In a closed cycle, fluid passes continuously round a closed circuit, through a thermodynamic cycle in which heat is received from a source at higher temperature, and heat rejected to a sink at low temperature and work output is delivered usually to drive an electric generator.

A gas turbine power plant may simply operate on a closed circuit as shown in Fig. 3.23.

Most gas turbine plants operate in "open circuit", with an internal combustion system as shown in Fig. 3.24. Air fuel cross the single control surface into the compressor and combustion chamber, respectively, and combustion products leave the control surface after expansion through the turbine.

The classical combined cycle for power production in a gas turbine and steam plant is normally associated with the names of Brayton and Rankine, respectively.

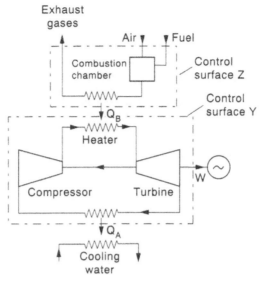

Fig. 3.23 Closed circuit gas turbine plant.

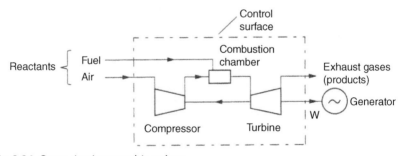

Fig. 3.24 Open circuit gas turbine plant.

Fig. 3.25 is simple representation of CCGT system. It demonstrates the fact that a CCGT system is two heat engines in series. The upper engine is the gas turbine. The gas turbine exhaust is the input to the lower engine (a steam turbine). The steam turbine exhausts heat to a circulating water system that cools the steam condenser.

An approximate combined-cycle efficiency (ηCC) is given as

$$\eta_{CC} = \eta_B + \eta_R - (\eta_B \times \eta_R) \tag{3.5}$$

Eq. (3.4) states that the sum of the individual efficiencies minus the product of the individual efficiencies equals the combine cycle efficiency. This simple equation gives significant insight to why combine cycle systems are successful.

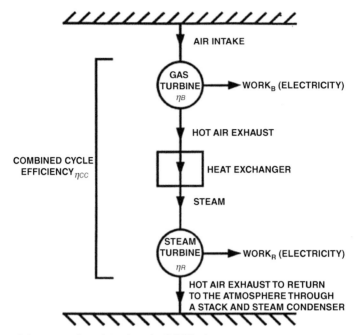

Fig. 3.25 Schematic of combined-cycle (CCGT) plant.

For example, suppose the gas turbines efficiency (Brayton) ηB is 40% (a reasonable value for a today's gas turbines) and that the steam turbine efficiency (Rankine) ηR is 30% (a reasonable value for a Rankine cycle steam turbine).

Utilizing Eq. (3.4) would lead to the following conclusion.

$\eta_{CC} = 0.4 + 0.3 - (0.4 \times 0.3)$
$\eta_{CC} = 0.58$
$\eta_{CC} = 58\%$

The combined-cycle efficiency of 58% is much greater than either the gas turbine or the steam turbines efficiencies separately. The 58% value is slightly misleading in that system losses were ignored. However, efficiency values in the 60% range have been recorded for CCGT systems in the past few years [6].

CCGT power plants come in many different configurations. Some companies choose to treat the gas turbine exhaust bypass stack as a commodity; others choose to incorporate a diverter damper into the turbine exhaust gas path. The diverter damper allows for the rapid configuration of the power plant as a combined cycle or simple cycle system. The initial cost of the diverter damper is much higher than the cost of treating the gas turbine exhaust stack as a commodity. However, the diverter damper allows for the gas turbines to be operated in simple cycle when HRSG or steam turbine repair or maintenance is required.

3.8 Advanced Computational Materials Proposed for GEN IV Systems

A renewed interest in nuclear reactor technology has developed in recent years, in part as a result of international interest in sources of energy that do not produce CO_2 as a byproduct. One result of this interest was the establishment of the GEN IV International Forum, which is a group of international governmental entities whose goal is facilitating bilateral and multilateral cooperation related to the development of new nuclear energy systems.

Historically, both the fusion and fission reactor programs have taken advantage of and built on research carried out by the other program. This leveraging can be expected to continue over the next 10 years as both experimental and modeling activities in support of the GEN-IV program grow substantially. The GEN-IV research will augment the fusion studies (and vice versa) in areas where similar materials and exposure conditions are of interest. However, in addition to the concerns that are common to both fusion and advanced fission reactor programs, designers of a future Deuterium-Tritium (DT) fusion reactor have the unique problem of anticipating the effects of the 14 MeV neutron source term. For example, advances in computing hardware and software should permit improved (and in some cases the first) descriptions of relevant properties in alloys based on *ab initio* calculations. Such calculations could provide the basis for realistic interatomic potentials for alloys, including alloy-He potentials that can be applied in classical molecular dynamics simulations. These potentials must have a more detailed description of many-body interactions than accounted for in the current generations which are generally based on a simple embedding function. In addition, the potentials used under fusion reactor conditions (very high pKa energies) should account for the effects of local electronic excitation and electronic energy loss. The computational cost of using more complex potentials also requires the next generation of massively parallel computers. New results of *ab initio* and atomistic calculations can be coupled with ongoing advances in kinetic and phase field models to dramatically improve predictions of the nonequilibrium, radiation-induced evolution in alloys with unstable microstructures. This includes phase stability and the effects of helium on each microstructural component.

However, for all its promise, computational materials science is still a house under construction. As such, the current reach of the science is limited. Theory and modeling can be used to develop understanding of known critical physical phenomena, and computer experiments can, and have been used to, identify new phenomena and mechanisms, and to aid in alloy design. However, it is questionable whether the science will be sufficiently mature in the foreseeable future to provide a rigorous scientific basis for predicting critical materials' properties, or for extrapolating well beyond the available validation database.

Two other issues remain even if the scientific questions appear to have been adequately answered. These are licensing and capital investment. Even a high degree of scientific confidence that a given alloy will perform as needed in a particular GEN-IV or fusion environment is not necessarily transferable to the reactor licensing or capital market regimes. The philosophy, codes, and standards employed for reactor licensing are properly conservative with respect to design data requirements.

Experience with the US Nuclear Regulatory Commission suggests that only modeling results that are strongly supported by relevant, prototypical data will have an impact on the licensing process. In a similar way, it is expected that investment on the scale required to build a fusion power plant (several billion dollars) could only be obtained if a very high level of confidence existed that the plant would operate long and safely enough to return the investment.

These latter two concerns appear to dictate that an experimental facility capable of generating a sufficient, if limited, body of design data under essentially prototypic conditions (i.e. with $\sim$14 MeV neutrons) will ultimately be required for the commercialization of fusion power. An aggressive theory and modeling effort will reduce the time and experimental investment required to develop the advanced materials that can perform in a DT fusion reactor environment. For example, the quantity of design data may be reduced to that required to confirm model predictions for key materials at critical exposure conditions. This will include some data at a substantial fraction of the anticipated end-of-life dose, which raises the issue of when such an experimental facility is required. Long lead times for construction of complex facilities, coupled with several years irradiation to reach the highest doses, imply that the decision to build any fusion-relevant irradiation facility must be made on the order of 10 years before the design data are needed.

Two related areas of research can be used as reference points for the expressed need to obtain experimental validation of model predictions. Among the lessons learned from Accelerated Strategic Computing Initiative (ASCI), the importance of code validation and verification has been emphasized at the workshops among the courtiers involved with such research.

Because of the significant challenges associated with structural materials applications in these advanced nuclear energy systems, the *Workshop on Advanced Computational Materials Science: Application to Fusion and Generation IV Fission Reactors* was convened by the US Department of Energy's Office of Science and the Office of Nuclear Energy, Science and Technology to ensure that research funded by these programs takes full advantage of ongoing advancements in computational science and the Department's investment in computational facilities. In particular, participants in the workshop were asked to:

1. examine the role of high-end computing in the prediction of materials behavior under the full spectrum of radiation, temperature, and mechanical loading conditions anticipated for advanced structural materials that are required for future GEN IV fission and fusion reactor environments, and
2. evaluate the potential for experimentally validated computational modeling and simulation to bridge the gap between data that are needed to support the design of these advanced nuclear technologies and both the available database and data that can be reasonably obtained in currently available irradiation facilities.

Like the requirements for advanced fusion reactors, the need to develop materials capable of performing in the severe operating environments expected in GEN IV reactors represents a significant challenge in materials science. There is a range of potential GEN-IV fission reactor design concepts and each concept has its own unique demands. Improved economic performance is a major goal of the GEN-IV designs. As a result, most designs call for significantly higher operating temperatures than the current generation of LWRs to obtain higher thermal efficiency. In many cases, the desired operating temperatures rule out the use of the structural alloys employed today. The very high operating temperature (up to 1000 °C) associated with the NGNP is a prime example of an attractive new system that will require the development of new structural materials (Table 3.3).

The operating temperatures, neutron exposure levels, and thermomechanical stresses for proposed GEN-IV fission reactors are huge technological challenges among material scientists and engineers. In addition, the transmutation products created in the structural materials by the high-energy neutrons produced in this generation of nuclear power reactors

Table 3.3 Structural materials.

System	Ferritic martensitic stainless steel alloys	Austenitic stainless steel alloys	Oxide dispersion strengthened steels	Ni-based alloys	Graphite	Refractory alloys	Ceramics
GFR	P	P	P	P		P	P
LFR	P	P	S			S	S
Molten Salt Reactor (MSR)				P	P	S	S
SFR	P	P	P				
SCWR-thermal spectrum	P	P	S	S			
SCWR-Fast spectrum	P	P	S	S			
VHTR	S			P	P	S	P

P = Primary, S = Secondary

can profoundly influence the microstructural evolution and mechanical behavior of these materials.

3.9 Material Classes Proposed for GEN IV Systems

The types of materials that were proposed in a DOE workshop in March of 2004 are tabulated as follows.

3.10 GEN IV Materials Challenges

A summary of these challenges for the next generation of NPPs are presented here. They are as follows:
- Higher temperature/larger temperature ranges
  - Examples
    - Very HTR (VHTR) coolant outlet temperature near 1000 °C.
    - Gas-cooled fast reactor (GFR) transient temperatures to 1600–1800 °C, gradient across core of ~400 °C.
    - FR to 800 °C steady-state outlet.
  - Issues
    - Creep
    - Fatigueness

- Toughness
- Corrosion/Stress corrosion cracking (SCC).
- Must drive modeling toward a predictive capability of materials properties in complex alloys across a wide temperature range.
- High flounce dose
  - Examples
    - LFR, Sodium-cooled fast reactor (SFR) cladding
    - Supercritical Water Reactor (SCWR) core barrel
    - GFR matrix
  - Issues
    - Swelling
    - Creep, stress relaxation
- Must drive modeling toward a predictive capability of materials properties in complex alloys to large radiation dose.
- Unique chemical environments
  - Examples
    - Pb and Pb-Bi eutectic.
    - Supercritical water.
    - High temperature oxidation in gas-cooled systems.
    - Molten salts.
  - Issues
    - Corrosion.
    - SCC/Irradiation-assisted stress corrosion cracking (IASCC).
    - Liquid metal embrittlement.
- Must drive modeling toward a predictive capability of chemical interactions in complex alloys to large radiation dose.

3.11 GEN IV Materials Fundamental Issues

The coevolution of all components of the microstructure, and their roles in the macroscopic response in terms of swelling, anisotropic growth, irradiation creep, and radiation-induced phase transformations should be studied within the science of complex systems. See Fig. 3.26.

In summary, we can conclude following:
- Six concepts have been identified with the potential to meet the GEN IV goals.
- Concepts operate in more challenging environments than current LWRs and significant material development challenges must be met for any of the GEN IV systems to be viable.

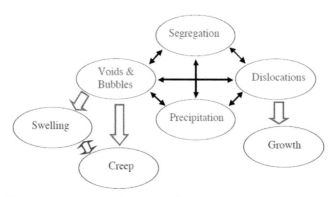

Fig. 3.26 Flow chart of materials fundamental issues.

- Experimental programs cannot cover the breadth of materials and irradiation conditions for the proposed GEN IV reactor designs
- Modeling and microstructural analysis can provide the basis for a material selection that is performed based on an incomplete experimental database and that requires considerable judgment to carry out the necessary interpolation and extrapolation

3.12 Capital Cost of Proposed GEN IV Reactors

Different PCS design trades may have substantial effects on the system capital cost. System optimization is typically complex, because, for example, increased PCS cost can increase cycle efficiency, reducing the reactor capital cost. The Generation IV Economic Modeling Working Group recommends two methodologies for modeling economics costs, a top-down method based on scaling and detailed information about similar systems, and a bottom-up method based on detailed accounting for all construction commodities, plant equipment, and labor-hours. For top-down methods, the EMWG recommends:

The first task is to develop a reference design to which cost-estimating techniques can be applied. The cost estimating part of this task generally is accomplished by considering the costs of equipment used for similar type projects and then scaling the equipment upwards or downwards. As an example, one might start cost estimating work on the VHTR by scaling reactor plant equipment from a project for which detailed estimates are available, such as the general atomics high temperature gas-cooled reactor (HTGR).

For the purpose of system comparison, the top-down method was adopted to estimate PCS parameters that are important in scaling relative capital costs.

The measures selected were those typically calculated to provide input for system cost estimates, and thus provide a basis for rough comparisons of system options. To provide an approximate baseline for comparison, where possible, comparisons were made with GEN II and GEN III+ LWR values. Fig. 3.26 shows such a comparison, quantifying steel and concrete inputs for the reference systems considered in the study. Several insights can be drawn from Fig. 3.26. For example, the 1500-MWe passive ESBWR LWR has slightly smaller inputs than the1970s LWRs, as well as the evolutionary Environmental Program Requirements (EPR).

But Fig. 3.27 also shows that it is possible to build high-temperature gas-cooled reactors, for example the 286 MWe GT-MHR, with smaller construction material inputs than for LWRs, due to the higher thermodynamic efficiency and power density. This shows that it is possible, with a high-temperature gas power cycle technology, to break the economic scaling of the large LWRs. This study also suggests that high-temperature; high efficiency gas-cycle power conversion can be adapted to other advanced reactor systems. For example, the even smaller inputs for the high-temperature, liquid-cooled, 1235-MWe AHTR-IT show that scaling economies may exist for HTRs. However, the material inputs for HTRs can be sensitive to equipment design choices and configurations, as shown by the differences

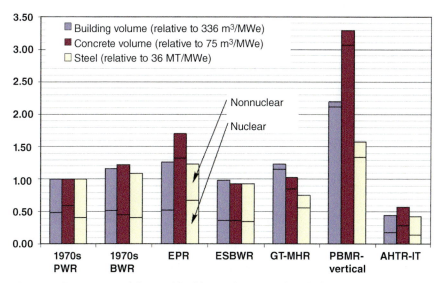

Fig. 3.27 Comparison of the total building volumes, and total plant steel and concrete inputs, for the reference HTR and LWR systems considered.

in Fig. 3.26 between the GT-MHR and the PBMR. Thus, careful attention to design tradeoffs is clearly important in the design of PCSs.

The selected capital costs that have been calculated for the reference systems in the study and are presented by the report from the UC Berkeley team [1] in Chapter 3 in more detail are based on the volumes of materials used:
- Structures costs:
  - Building volume ($m^3/MW(e)_{ave}$) (nuclear/nonnuclear)
  - Concrete volume ($m^3/MW(e)_{ave}$) (nuclear/nonnuclear)
- Reactor and PCS cost:
  - Reactor power density ($m^3/MW(e)_{ave}$)
  - PCS power density ($m^3/MW(e)_{ave}$) (nuclear/nonnuclear)
  - System specific steel ($MT/MW(e)_{ave}$) (nuclear/nonnuclear)
  - Turbomachinery–specific volume ($m^3/MW(e)_{ave}$)
  - System-specific helium ($kg/MW(e)_{ave}$) (nuclear/nonnuclear) (nonrenewable resource, correlates with building volume (blowdown))

For each of these figures of merit, the values for the nuclear and nonnuclear portions of the plant were estimated. This division recognizes the difference in costs for procuring and installing nuclear-grade materials. For example, for concrete and reinforcing steel, material costs are estimated to be 65% greater for nuclear-grade materials, and installation costs 30% greater

3.12.1 Economic and Technical of Combined-Cycle Performance

The output and efficiency of combined-cycle plants can be increased during the design phase by selecting the following features [18]:
- Higher steam pressure and temperature
- Multiple steam pressure levels
- Reheat cycles.

Additional factors are considered if there is a need for peak power production. They include gas turbine power augmentation by water or steam injection or a supplementary-fired HRSG. If peak power demands occur on hot summer days, gas turbine inlet evaporative cooling and chilling should be considered. Fuel heating is another technique that has been used to increase the efficiency of combined-cycle plants.

The ability of combined-cycle plants to generate additional power beyond their base capacity during peak periods has become an important design consideration. During the last decade, premiums were paid for power generated during the summer peak periods. The COE during the peak periods can be 70 times more expensive than off-peak periods. As the cost during the peak periods is much higher, most of the plant's profitability could be driven by the amount of power generated during these peak periods. Thus, plants that can generate large quantities of power during the peak periods can achieve the highest profits.

3.12.2 Economic Evaluation Technique

Plant output and efficiency are carefully considered during the initial plant design because they impact the COE in combination with fuel costs, plant capital cost, cost of capital and electricity sales. These factors will drive the gas turbine selection as well as the bottoming cycle design in combined-cycle operation.

As fuel costs increase, cycle selections typically include higher steam pressures, multiple steam pressure levels, reheat cycles, and higher steam temperatures. Once these selections have been made, other factors are addressed. Is there a need for peak power production with premiums paid for the resulting power? If so, gas turbine powers augmentation by way of water or steam injection or a supplementary-fired HRSG maybe the solution. Do peak power demands occur on a hot day (summer peaking)? This may suggest a potential benefit from some form of gas turbine inlet evaporative cooling or chilling [19].

For existing plants, some performance enhancement options can also be economically retrofitted to boost power output and efficiency. Although this research's primary focus is on options that enhance output, a brief discussion of fuel gas heating, which is a technique used to enhance combined-cycle plant efficiency, is provided.

The ability of utilities and independent power producers (IPPs) to generate additional power beyond a plant's base capacity during summer peak power demand periods has become an important consideration in the design of combined-cycle plant configurations. In recent years, utilities and IPPs within the United States have received premiums for power generation capacity during summer peak power demand periods. The price of electricity varies greatly as a function of annual operating hours. The variation is also highly region dependent. With price-duration curves that are sharply peaked, implying a few hours annually with very high rates, the majority of a plant's profitability could be driven by the high peak energy rates that can be achieved over a relatively short period of time. Thus, a plant that can economically dispatch a large quantity of additional power could realize the largest profits.

While current market trends should be considered during the design and development phase of a combined-cycle facility, forecasts of future market trends and expectations are equally important and warrant design considerations.

One of the primary challenges facing developers of new combined-cycle plants, as well as owner/operators of existing plants, is the optimization of plant revenue streams. As a result of escalating peak energy rates and peak demand duration, significant emphasis has been placed on developing plant designs that maximize peak power generation capacity while allowing for cost-effective, efficient operation of the plant during nonpeak power demand periods.

In addition to maximizing plant profitability in the face of today's marketplace, expectations of future market trends must be considered. Therefore, the goal is to determine which performance-enhancement options or combination of options can be applied to a new or existing combined-cycle plant to maximize total plant profits on a plant life-cycle basis.

With very few exceptions, the addition of power-enhancement techniques to a base plant configuration will impact base-load performance negatively and, hence, affect a plant's net revenue generating capability adversely during nonpeak periods [20].

In general, efficiency is the predominating economic driver during nonpeak generating periods, while capacity dominates the economic evaluation during peak power demand periods. Thus, it is extremely important to develop an economic model that considers both the COE during nonpeak periods while taking into consideration expectations of peak energy rates.

After having established baseline peak and nonpeak period performance levels for the various power-enhancement alternatives, a COE analysis technique is applied to determine alternatives that would afford the best overall life-cycle benefit. In addition to including both peak and nonpeak performance levels, the COE model includes the split between annual peak and nonpeak operating hours, the premium paid for peak power generation capacity, the cost of fuel, plant capital cost, the incremental capital cost of the enhancements, and the cost to operate and maintain the plant. This COE model then can be used to determine the sensitivity of a given power-enhancement alternative with respect to the economic parameters included within it [20].

Most peak power enhancement opportunities exist in the topping cycle (gas turbine) as opposed to the bottoming cycle (HRSG/steam turbine). In general, with the exception of duct firing within the HRSG, there are few independent design enhancements that can be made to a bottoming cycle that has already been fully optimized to achieve maximum plant performance. However, in general, performance enhancements to the gas turbines will carry with them an increase in bottoming cycle performance due to an associated increase in gas turbine exhaust energy [20].

3.12.3 Output Enhancement
The two major categories of plant output enhancements are:
1. gas turbine inlet air cooling and
2. power augmentation.

3.12.3.1 Gas Turbine Inlet Air Cooling
Industrial gas turbines operating at a constant speed have a constant volumetric flow rate. As the specific volume of air is directly proportional to temperature, cooler air has a higher mass flow rate. It generates more power in the turbine. Cooler air also requires less energy to be compressed to the same pressure as warmer air. Thus, gas turbines generate higher power output when the incoming air is cooler [21].

A gas turbine inlet air cooling system is a good option for applications where electricity prices increase during the warm months. It increases the power output by decreasing the temperature of the incoming air. In combined-cycle applications, it also results in improvement in thermal efficiency. A decrease in the inlet dry-bulb temperature by 10 °F (5.6 °C) will normally result in around a 2.7% power increase of a combined cycle using heavy-duty gas turbines. The output of simple-cycle gas turbines is also increased by the same amount.

Fig. 3.28 shows that a 10 °F (5.6 °C) reduction in gas turbine inlet dry-bulb temperature for heavy-duty gas turbines improves combined-cycle output by about 2.7%. The actual change is somewhat dependent on the method of steam turbine condenser cooling being used. Simple cycle output is improved by a similar percentage.

Several methods are available for reducing gas turbine inlet temperature. There are two basic systems currently available for inlet cooling. The first and perhaps the most widely accepted system is evaporative cooling. Evaporative coolers make use of the evaporation of water to reduce the gas turbine's inlet air temperature. The second system employs various ways to chill the inlet air. In this system, the cooling medium (usually chilled water) flows through an HX located in the inlet duct to remove heat from the inlet air.

Evaporative cooling is limited by wet-bulb temperature. Chilling, however, can cool the inlet air to temperatures that are lower than the wet bulb

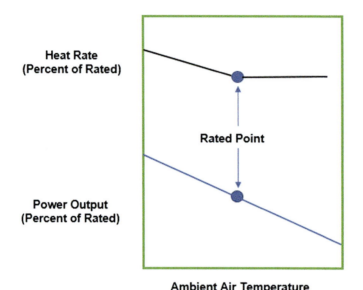

Fig. 3.28 Combined-cycle system performance variation with ambient air temperature. [19].

temperature, thus providing additional output although at a significantly higher cost.

Depending on the combustion and control system, evaporative cooling may reduce NOx emissions; however, there is very little benefit to be gained from current dry low NOx technology. This is another avenue that requires further analysis and investigation as well as collaboration between scientific communities, national laboratories, and industries.

3.12.3.2 Power Augmentation
Three basic methods are available for power augmentation:
1. Gas turbine steam/water injection
2. Supplementary-fired HRSG
3. Peak firing,

These are the three methods that General Electric is suggesting, and they need to be investigated further by nuclear power manufacturers and the community involved with enhancing nuclear power energy efficiency [17] using combined-cycle technology.

Other aspects of the cost of producing electricity are generally expressed in US$/MWh or US cts/kWh, depending on following parameters [20]:
- Capital cost of the project.
- Fuel cost

- Operation and maintenance cost

The capital cost per unit of electricity for a given power plant depends on following elements:
- Investment cost
- Financing structure
- Interest rate and return on equity
- Load factor of the plant (or equivalent utilization time)

The investment costs are the sum of the following positions:
- Power plant contract prices(s)
- Interest during construction (depending upon the construction time)
- Owner's cost for the realization of the project (project manager, owner's engineer, land cost, etc.)

The financing structure is defined by the debt-to-equity ratio of the financing and the return on equity is the return expected by the investors on their capital. Both are linked to the risks of the project.

The load factor results from the type of application the plant is intended for: Base, intermediate or peak load operation, and the availability and reliability of the power station.

Fuel costs per unit of electricity are proportional to the specific price of the fuel, and inversely proportional to the average electrical efficiency of the installation. This average electrical efficiency must not be mixed up with the electrical efficiency at a rated load. It is defined as follows:

$$\bar{\eta} = \eta \cdot \eta_{Oper} \tag{3.6}$$

where:

η is the electrical net efficiency at a rated load. (This is the percentage of the fuel that is converted into electricity at rated load for a new and clean condition)

η Oper is the operating efficiency, which takes into account the following losses:
- Start-up and shutdown losses
- Higher fuel consumption for part load operation
- Aging and fouling of the plant.

3.13 Combined-Cycle PCS Driven GEN IV Nuclear Plant

In this section, we review results of modeling a combined-cycle Brayton-Rankine PCS that are presented. The Rankine bottoming cycle appears to offer significant advantages over the recuperated Brayton cycle. The overall cycle was optimized as a unit and lower pressure Rankine systems seem to be more efficient. The combined cycle requires a lot less circulating water for a heat dump than current power plants. The modeling of thermal-hydraulic analysis was done based on steady state rather than transient state for the simplicity of the model, with help of computer and codes developed around it by this author and his coauthor, Patrick J. McDaniel at university of New Mexico, Nuclear Engineering Department around 2012.

A number of technologies are being investigated for the NGNP that will produce heated fluids at significantly higher temperatures than current generation power plants. The higher temperatures offer the opportunity to significantly improve the thermodynamic efficiency of the energy conversion cycle.

One of the concepts currently under study is the Molten Salt Reactor. The coolant from the Molten Salt Reactor may be available at temperatures as high as 800–1000 °C. At these temperatures, an open Brayton cycle combined with and Rankine bottoming cycle appears to have some strong advantages. Thermodynamic efficiencies approaching 50% appear possible. Requirements for circulating cooling water will be significantly reduced.

However, to realistically estimate the efficiencies achievable it is essential to have good models for the HXs involved as well as the appropriate turbomachinery. This study has concentrated on modeling all power conversion equipment from the fluid exiting the reactor to the energy releases to the environment.

3.13.1 Modeling the Brayton Cycle

Any external combustion or heat engine system is always at a disadvantage to an internal combustion system. The internal combustion systems used in current jet engine and gas turbine power systems can operate at very high temperatures in the fluid, and cool the structures containing the fluid to achieve high thermodynamic efficiencies. In an external energy generation system, like a reactor powered one, all of the components from the core to the HXs heating the working fluid must operate at a higher temperature than the fluid. This severely limits the peak cycle temperature compared to an internal combustion system. This liability can be overcome to a certain extent by using multiple expansion turbines and designing highly efficient HXs to heat the working fluid between expansion processes. Typically, the combustion chamber in a gas turbine involves a pressure drop of 3–5% of the total pressure. Efficient liquid salt to air HXs can theoretically be designed with a pressure drop of less than 1%. This allows three to five expansion cycles to achieve a pressure drop comparable to a combustion system. Multiple turbines operating at different pressures have been common in steam power plants for a number of years. In this study three to five gas turbines operating on a common shaft were considered. Multiple expansion turbines allow a larger fraction of the heat input to be provided near the peak temperature of the cycle significantly improving the thermodynamic efficiency. The exhaust from the last turbine is provided to the HRSG to produce the steam used in the Rankine bottoming cycle. The hot air after it passes through the HRSG is exhausted to the atmosphere. A detailed comparison of this system was made with a recuperated standalone Brayton cycle and the dual cycle appears to be more efficient for open systems.

3.13.2 Modeling the Rankine Cycle

The Rankine cycle was modeled with the standard set of components including the heat recovery steam generators (HSRG), a steam turbine, condenser, and high-pressure pump. Multiple reheat processes were considered. There is a slight efficiency advantage to include two reheat processes as per fairly standard design practices in today's power plants. The major limitation on the size of the steam system is the enthalpy available from high temperature air above the pinch point where the high-pressure water working fluid starts to vaporize. Below this point, there is still a significant enthalpy in the air which is readily available to heat the high-pressure water. There does not appear to be an advantage to including feedwater heaters in the cycle to bring the high-pressure water up to the saturation point. The possibility that an intercooler could be inserted between the two stages of a split compressor was considered. The cooling fluid for the intercooler was the high-pressure water coming out of the water pump. This process would combine the function of the traditional intercooler with the preheating of a typical feedwater heater. The effect of this addition to the two cycles had a marginal effect on the overall system efficiency and likely is not worth the cost, or effort, to implement.

3.13.3 Results

Multiple turbines and highly efficient HXs can produce highly efficient systems. A typical example for a turbine inlet temperature of 800 °C (1073 K) and a 100 MWe system using four gas turbine expansion processes and three steam turbine expansion processes produces a system with the following characteristics (Table 3.4).

Note that the bottoming cycle produces 40% of the power and has a mass flow rate that is approximately 10% of the air mass flow rate. Also, the heat rejected to cooling water is 53 MW for a 100 MW power plant. This

Table 3.4 System characteristic of the model.

Air turbine inlet temp	1073.0 K
Air turbine exit temp	903.0 K
Compressor ratio	21.8
Brayton(air)	205.6 kg/s
Brayton efficiency	31.3 %
Brayton power	60.2 MW
Rankine (water)	21.2 kg/s
Rankine efficiency	42.5 %
Rankine power	39.8 MW
Overall efficiency	51.7%
Heat rejection to water	53 MW

is roughly one-fourth of the heat rejection requirement for a current 33% efficient power plant operating at the same electrical power level. It may be possible to increase the condenser temperature for the bottoming cycle, at a decrease in overall efficiency, so that the condenser could be cooled by air. This would mean that this type of power plant would be free of the requirement to have a body of water nearby.

At a more modest turbine inlet temperature of 660 °C, a comparison of possible turbine exit temperatures and bottoming cycle pressures was made to determine the optimum overall efficiency as a function of these two parameters. The results are displayed in Fig. 3.29.

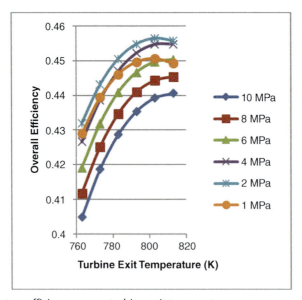

Fig. 3.29 System efficiency versus turbine exit temperature.

The optimum overall efficiency occurs at a turbine exit temperature of 803 K and a steam pressure of 2 Mega-Pascals. Higher Rankine cycle pressures might produce higher cycle efficiencies, but the mass flow rate must be lower due to the narrowing of the range between the turbine exit temperature and the water saturation temperature at the higher pressures. Thus, the combined cycle optimizes at a lower steam pressure than might be expected.

Additional work will be aimed at varying the expansion ratio per turbine and the penalties associated with air cooling the condenser. A more accurate thermal hydraulic analysis based on transient approach and further modeling should be taken under consideration, where finite elements and finite differences should be utilized for a better results. The above model used a FORTRAN computer code on Window/PC.

We also, suggest reading Appendix D and Appendix E of this book as well.

References

[1] P. Peterson, H. Zhao, R. Ballinger, R. Fuller, M. Forsha, B. Nichols, C. Oh and M.E. Vernon, Next generation nuclear plant power conversion study: Technology options assessment. September 1, 2004.

[2] B. Zohuri, P. McDaniel, Combined Cycle Driven Efficiency for Next Generation Nuclear Power Plants, An Innovative Design Approach, second ed, Springer Publishing Company, New York, 2014.

[3] B. Zohuri, P.J. McDaniel, Thermodynamics in Nuclear Power Plant Systems, Springer Publishing Company, New York, 2014.

[4] B. Zohuri, Thermal-Hydraulic Analysis of Nuclear Reactors, second ed, Springer Publishing Company, New York, 2017.

[5] M.N Ozisik, Heat Transfer: A Basic Approach, McGraw-Hill, NY, 1985.

[6] F.P. Incropera, D.P. DeWitt., Fundamentals of Heat and Mass Transfer, third ed, Wiley, New York, 1990, pp. 658–660.

[7] F.P. Incropera, D.P. DeWitt, T.L. Bergman, A.S. Lavine, Fundamentals of Heat and Mass Transfer, sixth ed, John Wiley & Sons, NJ, 2006, pp. 686–688.

[8] B. Zohuri, P.J. McDaniel, C. de Olivera, Air Brayton cycles for nuclear power plants. In: Application of Compact Heat Exchangers For Combined Cycle Driven Efficiency In Next Generation Nuclear Power Plants, Springer, Cham, 2016, pp. 103–124.

[9] B. Zohuri, P.J. McDaniel, C. de Olivera, A Comparison of a recuperated open cycle (air) Brayton power conversion system with the traditional steam Rankine cycle for the next generation nuclear power plant, In Proceeding of ANS Transactions, 2014 June 2014.

[10] P.J. McDaniel, B. Zohuri, C. de Olivera, A Combined cycle power conversion system for small modular LMFBRs, Small Modular Reactors as Renewable Energy Sources, Springer (2019).

[11] P.J. McDaniel, B. Zohuri, C. de Oliveira, J. Cole, A combined cycle power conversion system for the next generation nuclear power plant, Proceedings of ANS Transactions, 2012 2012 November.

[12] http://energyfromthorium.com/2014/04/04/closed-loop-brayton-cycle-sandi5-national-laboratory/.

[13] J. Pasch, T. Conboy, D. Fleming, and G. Rochau, Supercritical CO2 recompression Brayton Cycle: Completed assembly description, SANDIA REPORT SAND2012-9546 October 2012. https://prod-ng.sandia.gov/techlib-noauth/access-control.cgi/2012/129546.pdf.
[14] http://www.netl.doe.gov/publications/proceedings/11/utsr/pdf/wed/Wright%20SCO2%20Power%20Cycle%20Summary%20UTSR%202011%20v2a.pdf.
[15] W.M Kays, A.L. London, Compact Heat Exchangers, third ed, McGraw-Hill, New York, 1984.
[16] K.J. Bell, S. Mochizuki, R.K. Shah, V.V. Wadekar, Compact Heat Exchangers for the Process Industries, Begell House, Inc, New York, 1997.
[17] C. Jones and J. Jacob, III, Economic and technical considerations for combined-cycle. https://www.ge.com/content/dam/gepower-pgdp/global/en_US/documents/technical/ger/ger-4200-eco-tech-considerations-for-cc-performance-enhancement.pdf.
[18] LS. Langston, G. Opdyke, Introduction to Gas Turbine for Non-Engineers, Vol. 37, Global Gas Turbine News, (1997). No. 2.
[19] Chuck Jones and John A, I.I.I. Jacobs, Economic and Technical Considerations for Combined-Cycle Performance-Enhancement Options, GE Power Systems Schenectady, NY. GER-4200.
[20] K. Rolf, F. Hannemann, F. Stirnimann, B. Rukes, Combined-Cycle Gas & Steam Turbine Power Plants, third ed, PennWell Publication, 2009.
[21] C. Jones, J. Jacob III, Economic and Technical Considerations for Combined-Cycle Performance-Enhancement Options, GE Power Systems, 2000 GER-4200October.

CHAPTER 4

Advanced Power Conversion System Driven by Small Modular Reactors

4.1 Introduction

Currently in the United States there are less than 100 reactors operating [1]. The number of operating reactors peaked in the late 1990s and has started to decrease since. There are a number of reasons for this but most of them are related to economic competition. The availability of cheap natural gas is the most obvious one. Part of the problem is the allocation of costs to nuclear generated electricity. In the United States only approximately 10% of the cost of nuclear electricity is the cost of fuel. Over 50% of the cost is the capital cost of the plant, including financing during construction. The last build of nuclear reactors in the United States centered on 1 GW plants. The conventional wisdom said that the plants had to be this big to take advantage of economies of scale. However, since the great build of nuclear power plants in the 70s and 80s, most utilities have added capabilities at the 100 MW(e) to 500 MW(e) level. The demand for electricity has continued to grow, but it has been more economical to keep up with demand by building smaller power plants more rapidly. This leads to lower finance charges and faster cost recovery from an operating plant. Of course, this easily applies to plants burning natural gas, or coal. Regulatory delays are another competition impediment for nuclear plants as there is not a Fossil Regulatory Commission that oversees every element of construction leading to its own set of construction delays.

The construction and operational safety instilled in the US nuclear industry by the Nuclear Regulatory Commission is essential to its survival. It is essential to achieving the benefits of nuclear generated electricity for the American people. The demands of this type of regulatory body must be part of any advanced nuclear power plant design. To speed up the construction and licensing process two things can be implemented and still maintain safety excellence. First the design of a given type of power plant should be standardized, and second the plant itself should be as independent of siting characteristics as possible. Neither of these are new ideas, but Small Modular Reactors (SMRs) in many ways make them easier to implement. Significant reductions in the time required for licensing can provide an incentive for utilities to consider new nuclear power plants more favorably.

The second major problem faced by a stressed nuclear industry is the lack of manufacturing capability. During the major build of the 70s and 80s there were four large competing firms building nuclear power plants in the US. Westinghouse had been around since the birth of commercial nuclear power at a shipping port. General Electric had developed the boiling water reactor as a major competitor to the pressurized water reactors (PWR) following the nuclear submarine designs. Babcock and Wilcox had developed their own version of the PWR as had Combustion Engineering. Today Combustion Engineering is out of the business as a nuclear system provider. General Electric (GE) incorporation has lost their major vessel manufacturing capabilities and seeded that capability to Hitachi. Westinghouse has filed for bankruptcy has become an international company with most of their business overseas. Babcock and Wilcox has focused on their support for the nuclear navy and has not taken an aggressive commercial stance since the Three Mile Island accident. It is not clear that there is a foundry in the United States that can produce pressure vessels for a 1 GW plant. Certainly, there would not be enough business producing 1 GW vessels to keep a foundry afloat if the only market were the US electric industry. SMRs offer the opportunity for smaller vessels and more of them.

The next major shift in the electrical energy market faced by advanced nuclear power plants is the onset of extensive renewable energy systems. Solar and Wind systems are increasing at a phenomenal rate. From an economic standpoint, they have some of the same characteristics as nuclear. They are very capital intensive with a very low cost, or negligible cost, for their source of energy (Fig. 4.1). However, once again there is not a Renewable Regulatory Agency to prescribe construction standards. Thus, capital construction is not as uncertain as similar nuclear installations. The real problem they present for the nuclear grid is that they do not control their sources of energy. This makes them intermittent and not easily capable of matching output to demand. At the rate that renewable systems are being brought into the marketplace, they have saturated the demand for electricity at certain times and certain periods of the year. The infamous "duck curve" has been observed in the California market by several organizations [2].

Fig. 4.1 The California duck curve.

The Day-Ahead Prices for 2017 appear to form the outline of a duck. Note that for a significant part of the day the prices are negative. That is because the solar energy systems are supplemented financially and someone besides the customers are willing to pay for the generation capacity. A similar problem occurs in Germany with regard to wind energy in the winter on the North Sea. Several of Germany's neighbors get free electricity part of the months due to its over production when strong winds occur. There are also places in the United States where wind energy saturates the electricity market for short periods. The obvious answer to this intermittent over production is some kind of storage. Unfortunately, it is very expensive to store electrical energy at this time and a storage capability adds to the capital investment of a solar or wind power station [2].

Another challenge to new nuclear builds is the lack of cooling water to get rid of the waste heat required by current thermodynamic cycles. Typical nuclear plants have efficiencies in the range of 33%–35%. This means that they must get rid of 67%– 65% of the energy produced. They do this by heating environmental water or vaporizing water in the cooling towers that have become the symbols of nuclear plants. Conventional coal and gas plants are slightly more efficient, but still generally reject more energy than they produce in terms of electricity [3,4]. The latest combined cycle (CC) plants do achieve efficiencies approaching 60% or better and are finally able to produce more electrical energy than the waste heat they have to dump. In all cases the waste heat goes into a circulating water system that either heats environmental water and then returns it to the environment to be cooled by atmospheric processes or evaporates it in a cooling tower. In either case the atmosphere becomes the ultimate heat sink. Currently slightly over 50% of the freshwater in the United States is used to cool power plants. Not all of the freshwater is consumed, but this is still an amazing statistic. Construction of a new power plant is limited by the requirement to have a water heat dump nearby. This is why, for instance, all of Japan's nuclear power plants are built near the coast. All of the US nuclear power plants embody a water heat dump. The Palo Verde power station in Arizona is limited from expanding due to a water shortage, though there is an increased electrical energy demand in its market area.

The final problem facing a new generation of nuclear power plants is the lack a waste repository for commercial spent nuclear fuel and a reprocessing capability to recover the plutonium and remaining U-235 in spent fuel. For a number of years, it was thought that the best way to use the US Uranium reserves would be to build a series of breeder reactors that would produce more nuclear fuel than they consumed. This definitely requires a reprocessing capability to recover the plutonium from the once through fuel. As the United States has forgone that capability for the near future, the next best thing is to achieve as high of a conversion efficiency as possible in the reactors that are built. Basically, the idea is to design the cores so as to burn the plutonium in place in the fuel elements that it was produced in. For a 3–5% enriched fuel element in a current light water reactor (LWR) at the end of its 3–4 year burn cycle, the bred plutonium is producing as much energy as the remaining U-235 in the element. If 19.75% enriched fuel elements are loaded in a fast reactor it is possible to extend the refueling cycle significantly and burn more of the plutonium that has been produced. Conversion ratios exceeding 0.9 may be achievable. A waste repository and hopefully a reprocessing capability will still be required but the magnitude of both can be significantly reduced by developing future reactors with very high conversion ratios.

In this Chapter and Chapters 5, 7, and 8 an SMR with a new power conversion system will be described that addresses most of these problems. This reactor will be a fast reactor that is either sodium, lead, or molten salt cooled and uses an Air-Brayton power conversion system. The Air-Brayton power conversion will consist of two types—a CC with a steam bottoming cycle, or a recuperated cycle based on air alone. As nothing comes to fruition in this modern age without a simple acronym, the Nuclear Air-Brayton CC (NACC) will be identified as an NACC system. The Nuclear Air-Brayton Recuperated Cycle system will be identified as an NARC system. Both acronyms are pronounceable which is even better.

4.2 Currently Proposed Power Conversion Systems for SMRs

In the 1950s when nuclear power was first being considered as a possible source of submarine propulsion and electric power generation, the state-of-the-art power conversion systems for power plants were all steam boilers. Thus, the earliest power plants and most that exist today use their nuclear heat to boil water. Some of the SMRs that have been proposed follow this technology and are simply smaller versions of the current generation of 1 GW LWRs. The SMR's proposed by Holtec and BWXT fall into this category.

Primarily with safety in mind, the NuScale reactor uses a natural convection water system to extract heat from the reactor core. This is a concept pioneered in nuclear submarines to reduce the noise of the reactor pumps for more stealthy operation. It is the most advanced SMR in terms of design and the licensing process. It currently estimates a thermodynamic efficiency of 31%, somewhat lower that current Generation II or III LWRs.

The newest power conversion system of interest for an SMR is the supercritical CO_2 system. The seminal report describing this technology for nuclear reactors was published in 2004 by V. Dostal, M. J. Driscoll and P. A. Hejzlar from MIT [5]. The concept was picked up by researchers at Sandia National Laboratories and is currently being developed there by department of energy (DOE)/nuclear energy (NE). The high pressures possible (~7 MPa) enable very small turbines and heat transfer equipment. Efficiency predictions will be discussed in comparison with Air-Brayton systems subsequently. Currently there are no SMRs advocating this technology as a baseline, but it is likely that it will be adopted for SMRs as it matures.

A fourth power conversion system that has been developed for current fossil steam systems is the Super Critical Water power conversion system. SMRs in the 30–150 MW(e) power class have been proposed with this power conversion system. Its main advantage is that the working fluid does not change phase. This requires the system to operate above the water critical point at 647 K and 22.1 MPa. Though this is more than double the pressure of conventional steam systems, the technology has well penetrated the fossil steam market.

Other possibilities for SMR power extraction and conversion systems are the gas-cooled reactors. These are not easily adapted to the Small part of an SMR as the gas heat extraction capability in the core requires a large surface area, implying large cores. Efficiencies of these systems will be compared with the proposed liquid metal/molten salt Air-Brayton systems proposed here, but the development of more compact cores will require significant time to catch up with the more compact systems.

4.3 Advanced Air-Brayton Power Conversion Systems

Advanced Air-Brayton power conversion systems are modeled after current-generation gas turbine systems and take advantage of much of the technology developed for these systems. At the time that nuclear power was being developed, the gas turbine was going through a rapid development period also. The development of jet-powered aircraft provided a very strong incentive for advances in gas turbine technology. This technology was adapted to stationary electric power plants to provide peaking power using kerosene as a fuel. The gas turbine plants could start much faster and shut down quicker than the massive steam boiler plants even if the fuel was more expensive. With the advent of cheap natural gas, baseload gas turbine planets came into their own. Then it was observed that the increased temperatures available to the gas turbines allowed their exhaust to be used to heat water in a conventional boiler and the CC was invented. Currently gas turbine CC (GTCC) power plants achieve efficiencies over 60%. The efficiency for a gas turbine plant is the driving performance measure because 85% of the cost of producing electricity is the cost of the fuel consumed.

A system diagram for a typical GTCC is provided in Fig. 4.2. The ambient air is taken in through the air compressor, combined with the fuel, and burned in the combustion chamber. Then it is expanded through the turbine, passed through the heat recovery steam generator (HSRG), and exhausted to the atmosphere. The bottoming steam cycle starts with liquid water at the entrance to the pump where it is raised to high pressure. It then passes through the HSRG where it is vaporized in much the same fashion that it would be in a conventional boiler. The steam is then expanded and

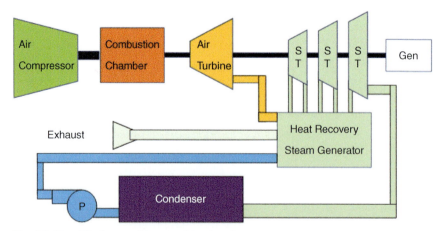

Fig. 4.2 Simplified gas turbine combined cycle system.

reheated through a series of typically three turbines. The exhaust goes to a condenser that extracts the waste heat and condenses the steam to water to start the cycle over again. Note that a circulating water system is required to extract the waste heat from the condenser and deposit it in the environment in some fashion.

This loop is not shown in Fig. 4.2. Also note that the system diagram is very simplified to present the major components only. Any real system would at least have several feed water heaters to improve the efficiency of the steam cycle slightly. It is also worth pointing out that the steam cycle is a closed cycle and the working fluid is used continuously. For the Air-Brayton cycle, the air is used once and exhausted to the environment.

The simplest NACC system looks exactly like the GTCC system except that a heat exchanger is substituted for the combustion chamber and another fluid loop is added to transfer the working fluid going through the reactor to this heat exchanger. Actually, for the systems to be considered here, the heat transfer fluid going through the reactor passes through another heat exchanger to a similar working fluid that is then passed through the heat exchanger that drives the Air-Brayton cycle. For a sodium-cooled reactor this means the primary sodium passes through the reactor and then through a sodium-to-sodium intermediate heat exchanger. The heated sodium then goes to the primary sodium-to-air heat exchanger that drives the air turbine.

The recuperated Air-Brayton system (NARC) simply replaces the HRSG with a heat exchanger that preheats the compressed air before it enters the primary heat exchanger and recovers some of the waste heat in the turbine exhaust. After passing through the recuperator heat exchanger, the exhaust is vented through a stack to the atmosphere. Thus, the waste heat from an NARC system is deposited in the atmosphere without going through the heating of a circulating water system.

It is useful to contrast the primary heat exchanger (sodium-to-air) with a combustion chamber. The heat exchanger cannot heat the working fluid to as high a temperature as the combustion chamber. In the heat exchanger the temperature change going from the solid material to the gas involves a temperature drop so the gas temperature must always be below the temperature of the solid heat exchanger material. In the combustion case the temperature drop is in the other direction where the gas is at a higher temperature than the combustion chamber material. As the combustion chamber can be cooled, this temperature drop could be quite significant. The gas temperature impinging on the turbine, the prime indicator of thermodynamic efficiency, will always be lower for the NACC or NARC systems than that for the GTCC system. There is a slight compensation though in that the pressure drop can be lower for the heat exchanger than for the combustion chamber. Typical pressure drops in combustion chambers are on the order of 3–5% whereas heat exchanger pressure drops can be designed to be less than 1%.

The other difference is that the NACC/NARC systems do not change the working fluid. Combustion systems use up the oxygen in the compressed air, and though they typically do not reach stoichiometric temperatures implying that all of the oxygen is burned, most of the oxygen is burned. For the NACC/NARC systems the air is only heated. Borrowing a trick from steam systems, the air can be expanded through the first turbine and then reheated and expanded again. For steam cycles, the steam is typically expanded through as many as three turbines. As the pressure drop through the primary heat exchanger can be as much as one-fifth that through a combustion chamber, it makes sense to consider as many as four reheats and five expansions of the air passing through the turbines.

With these thoughts in mind a typical system diagram for a two-turbine NACC system is presented in Fig. 4.3.

A typical system diagram for a two-turbine NARC system is presented in Fig. 4.4.

The two-turbine systems are the simplest multiturbine NACC/NARC systems that we will consider. Note that the air compressor, air turbines,

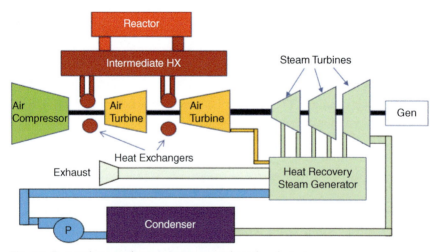

Fig. 4.3 Two-turbine nuclear air-Brayton combined cycle system.

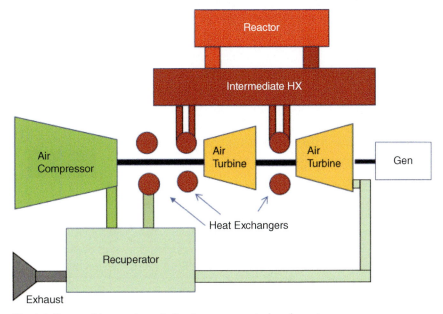

Fig. 4.4 Two-turbine nuclear air-Brayton recuperated cycle system.

steam turbines, and generator are all on the same shaft. Another configuration would have the air turbines and the steam turbines on different shafts. This would necessitate two smaller generators. For the discussion here the differences between these configurations will not be considered. However, one configuration change that is of interest is the conversion of the last air turbine to a power, or free, turbine. In this case the first turbine is connected to the air compressor shaft and it drives the compressor. The power, or free, turbine is not connected to this shaft, but is only connected to the generator. This is a common configuration in what are called turbo-shaft engines. The difference that will be considered here is that the working fluid (Air) will pass through another heat exchanger prior to entering the power turbine. In combustion systems this is not done because the air cannot be re-burned.

Additionally, it is possible to add a recuperator to the NACC system after the air exits the HRSG and before it is exhausted. This is not done with standard GTCC systems but will be considered here to improve a near-term system's efficiency.

Before getting into the design and analysis of components and cycles it is worth pointing out that other CCs have been proposed. The steam cycle has been replaced by an organic cycle in at least one design where the organic fluid chosen is toluene. As the bottoming cycle is a closed cycle, water is not unique as a working fluid and the organic cycle has some advantages.

Another CC has been proposed for space power plants that has two closed cycles. The topping cycle is a mixture of helium and xenon and the bottoming cycle uses isobutane as the working fluid. This cycle appears to be significantly more efficient than other proposed power conversion cycles, which for space power significantly reduces the size of the radiator used to dump waste heat.

4.4 Design Equations and Design Parameters

Two designs will be developed, one for near-term systems with sodium cooling, and one for advanced systems with molten salt cooling. These systems are representative of what a 50 MW(e) SMR system based on an Air-Brayton power conversion might look like.

4.4.1 Reactors

It would seem that a book on SMRs would go into a detailed design of the reactor core and heat removal system. However, the reactors of interest here are all of the liquid metal–type—sodium cooled, lead cooled, lead-bismuth cooled, and molten salt cooled. (Note the molten salt liquid fuel is not being considered as sizing has not been established very definitively.) Many prototypes have been built and their characteristics are well documented in the book by Waltar et al. [6].

A very simple approach is taken to the nuclear core for these systems. A pool-type liquid metal arrangement is assumed, and a simple linear regression is fitted to the power level versus size data from Waltar et al. based on the liquid metal systems that have been built around the world. The fitted curve is presented in Fig. 4.5.

This is a somewhat crude approach but is based on actual experience. Certainly, it would be desirable to beat the curve, but the curve is used only to show a comparison between the reactor part and the power conversion part of the plant. The other parameter that is of interest for fitting the rector into the modeling here is the temperature of the hot air at the entrance to the first turbine.

Current technology for sodium, and possibly lead or lead-bismuth systems, seems capable of achieving a temperature of 510 °C or 783 K. Molten salt systems are expected to be able to achieve higher temperatures so a

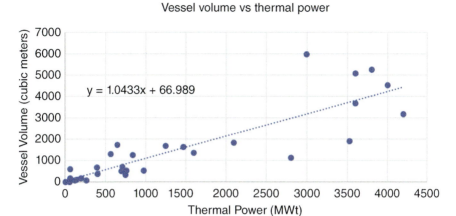

Fig. 4.5 Estimated volume versus reactor thermal power for liquid metal reactors.

temperature for them will be estimated at 675 °C or 948 K. The following analysis will look at near-term possible temperatures of 783 K and developmental temperatures of 948 K. High temperature gas-cooled reactors (HTGR) theoretically can reach temperatures much higher than these but analyzing an HTGR is beyond the goals here. The vessel volume will then be estimated at 1.04 times the thermal power of the reactor in megawatts plus 67 cubic meters. A near-term SMR reactor will deliver hot air to the first turbine at 783 K and an advanced SMR will deliver hot air to the first turbine at 948 K.

4.4.2 Air Compressors and Turbines

For all of the analyses to follow the concept of a "rubber" engine applies. That is, the equipment is rebuilt every time to match the desired conditions. This is different from considering a given "solid" engine and looking at its performance under different conditions. As the market for gas turbines is so large, the approach taken here is to look at the desired characteristics of a component, and then see if any readily available components meet that requirement. Even if they do not meet it exactly, if they are close, it may be worthwhile to sacrifice some performance for a readily available developed component. Thus, each component is designed to optimize the performance of the specific power plant considered.

With this in mind, the basic requirement for a compressor is to increase the pressure of the working fluid. When it does this it also heats the working fluid. The heating of the working fluid defines the work required to drive the compressor. The defining equations are:

$$T_{out} = T_{in} CPR^{\left(\frac{\gamma-1}{\gamma e_c}\right)} \quad (4.1)$$

where
CPR = compressor pressure rate
e_c = the polytropic efficiency for the compressor
γ = ratio of specific heats for air

$$W_{comp} = C_p (T_{out} - T_{in}) \quad (4.2)$$

where
C_p = air constant pressure specific heat

Note all calculations follow the standard practice of performing analyses on a per unit mass basis. The efficiency of turbines and compressors will be calculated using a quantity called polytropic, or small stage, efficiency. This efficiency is taken as independent of the pressure ratio. This allows the comparison of performance across multiple pressure ratios. It is also worth pointing out that the specific heat and ratio of specific heats are not constant for air, so an average value must be chosen based on the two temperatures at the start and finish of a process. This value is solved for iteratively by estimating the final temperature and then updating the specific heat and ratio of specific heats until all converge.

The classic thermodynamic efficiency can be calculated for a compressor or turbine by calculating an ideal temperature that would be produced with a polytropic efficiency of 1.0 and then comparing the two temperature changes. For instance, for a compressor we would have:

$$T_{out,ideal} = T_{in} CPR^{\left(\frac{\gamma-1}{\gamma}\right)}$$

and

$$e_{th} = \frac{W_{ideal}}{W_{actual}} = \frac{T_{out,ideal} - T_{in}}{T_{out} - T_{in}} \quad (4.3)$$

The governing equations for a turbine are

$$T_{out} = T_{in} \left(\frac{1.0}{CPR}\right)^{\frac{(\gamma-1)e_t}{\gamma}} \quad (4.4)$$

where
e_t = polytropic efficiency for the turbine
and

$$W_t = C_p (T_{in} - T_{out}) \quad (4.5)$$

In addition to estimating the thermodynamic performance for a particular power conversion system, an attempt will be made to estimate the size of the components, particularly the heat exchangers. To estimate the sizes for the compressors and turbines the actual mass flow must be calculated. This of course will depend on the power generated. The net electrical power generated by the Air-Brayton part of the cycle will be estimated as:

$$P(e)_{net} = \dot{m}\sum_i W_{t,i} - \dot{m}\sum_j W_{c,j} \qquad (4.6)$$

This equation is then solved for the mass flow rate given the desired electrical power from the Air-Brayton part of the cycle.

The compressors are sized based on-5-hub to tip ratio of 0.5, with a blade solidity of 0.05, and an entering Mach number of 0.4. This is enough information to calculate a compressor radius. A pressure ratio increase per stage is set at 1.25. Then the length of the blades is calculated based on the density entering a given stage for a constant axial velocity. The width of the stage is then calculated as 0.784 times the height of the blades. This gives a length for the compressor, and subsequently a volume.

The air turbine sizes are estimated in much the same way with a hub-to-tip ratio of 0.7, a blade solidity of 0.05, an exit Mach number of 0.3, and a pressure drop factor per stage of 2.5. There are many possible designs for compressors and turbines, but these parameters seem to be about average for current designs. The volumes of compressors and turbines are not major contributors to the overall system volume in the end. Thus, these approximations seem reasonably adequate.

Perhaps the most critical dimension estimated is the radius of the compressor as that is used to estimate the polytropic efficiency based on a correlation developed by Wilson and Korakianitis [7].

$$e_c = 0.862 + 0.015\ln(\dot{m}) - 0.0053\ln(r_c) \qquad (4.7)$$

where
$\dot{m}$ = the mass flow rate in kg/s
r_c = compressor pressure ratio

Wilson et al. [7] also developed a similar correlation for the polytropic efficiency of air turbines given by

$$e_t = 0.7127 + 0.03\ln(d_m) - 0.0093\ln(1/r_t) \qquad (4.8)$$

where
d_m = rotor mean diameter in mm
r_t = pressure ratio for the turbine

The sensitivity of the overall system size will not be addressed for air compressor and turbine performance as they are a small part of the volume of the overall systems. Steam turbine systems are modeled similar to air turbines except that a simple thermodynamic efficiency of 0.95 is assumed in all cases. This is in the range of recent state-of-the-art turbines [8,9]. A hub-to-tip ratio of 0.6, a blade solidity of 0.05, and an exit Mach number of 0.3 are used to compute the turbine diameter. A pressure drop factor per stage of 1.5 and a stage width proportional to 0.4 times the blade height were used to estimate the turbine length.

4.4.3 Heat Exchanger

There are numerous heat exchangers in each of the systems to follow. In many cases they make up the largest fraction of the system volume. The largest component for the NACC system is generally the HSRG which includes an economizer, evaporator, and generally three superheaters. For the NARC systems the largest component is generally the recuperator as it is an air-to-air heat exchanger. The heat transfer in each heat exchanger is calculated by the classic equation

$$Eff = \frac{C_i(T_{max,i} - T_{min,i})}{C_{min}(T_{hot,in} - T_{cold,in})} \qquad (4.9)$$

where

Eff = heat exchanger efficiency

C_{min} = minimum mass flow rate times specific heat for the two fluids

$T_{hoy,in}$ = the temperature of the hot fluid entering the heat exchanger

$T_{cold,in}$ = the temperature of the cold fluid entering the heat exchanger

C_i = the mass flow rate times the specific heat for the ith fluid, either hot or cold

$T_{max,i}$ = the maximum temperature for the ith fluid

$T_{min,in}$ = the minimum temperature for the ith fluid

The baseline efficiency for all heat exchangers is assumed to be 0.95. A pressure drop of 1.0% is also assumed for both fluids. Rather than using pressure drops, the parameter of interest will be defined as the pressure ratio. A 1% pressure drop means the exit pressure from the heat exchanger is 99% of the inlet pressure. The size of the heat exchanger can then be calculated with at least one fluid achieving the 1% pressure drop. This applies to all heat exchangers, be they liquid metal, or molten salt-to-air, or air-to-air, or air-to-water, or air-to-steam. The overall size to achieve a 0.95 efficiency also depends on the flow path within the heat exchangers, the ratio of the minimum-to-maximum mass flow rates times heat capacities, and the number of heat transfer units for a specific design. Flow paths can be counter-flow or cross-flow. There are several varieties of cross-flow of interest. Transfer units are defined as

$$N_{tu} = \frac{AU}{C_{min}} \qquad (4.10)$$

where
A = Heat transfer area
C_{min} = Is defined before and U is given by the following equation as

$$\frac{1}{U} = \frac{1}{\eta_h h_h} + \frac{1}{\frac{k}{a}} + \frac{1}{\eta_c h_c} \qquad (4.11)$$

where
h_h, h_c = heat transfer coefficient for the hot and cold surfaces,
h_h, h_c = surface efficiencies for hot and cold surfaces (fins etc.),
k/a = ratio of wall thermal conductivity to its thickness.

And the hot surface area has been assumed to be the same size as the cold surface area and the wall surface area. If that is not the case, then an average must be taken. Then the functional relationship is:

$$\mathit{Eff} = f(N_{tu}, C_{min}/C_{max}, \mathrm{flowpath}) \qquad (4.12)$$

These functional relationships have been taken from the text by Kays and London [10]. For the systems considered here the following configurations taken from Kays and London will be used.

4.4.3.1 Primary Heat Exchangers—Sodium-to-Air, Molten Salt-to-Air
Type: cross-flow unmixed
Surfaces: Louvered plate-fin
Pitch: 437 per meter
Heat transfer area per volume: 1204 m^2/m^3

4.4.3.2 Economizer—Air to Water
Type: cross-flow unmixed
Surfaces: Louvered plate-fin
Pitch: 437 per meter
Heat transfer area per volume: 1204 m^2/m^3

4.4.3.3 Superheaters—Air to Steam
Type: Cross-flow unmixed
Surfaces: Louvered plate-fin
Pitch: 437 per meter
Heat transfer area per volume: 1204 m^2/m^3

4.4.3.4 Condenser—Steam to Water
Type: Cross-flow unmixed
Surfaces: Louvered plate-fin
Pitch: 437 per meter
Heat transfer area per volume: 1204 m^2/m^3

4.4.3.5 Recuperator—Air to Air
Type: cross-flow unmixed
Surfaces: Plate-fin
Pitch: 1789 per meter
Heat transfer area per volume: 4372 m^2/m^3

4.4.3.6 Intercooler—Water to Air
Type: Cross-flow unmixed
Surfaces: Plate-fin
Pitch: 1789 per meter
Heat transfer area per volume: 4372 m^2/m^3

4.4.4 Pumps and Generators

The efficiency for water pumps was simply assumed to be 80% and the generators were assumed to be 99% efficient. A mechanical efficiency of 99% was assumed to account for frictional losses in the turbocompressors.

4.4.5 Connections and Uncertainty

The size of components was increased to allow for uncertainty and connections. The increase factors are as follows.

Reactor—estimated volume = 1.00 × calculated volume
Compressors—estimated volume = 1.20 × calculated volume
Air turbines—estimated volume = 1.10 × calculated volume
Primary HXs—estimated volume = 1.10 × calculated volume
Superheater—estimated volume = 1.20 × calculated volume
Steam turbines—estimated volume = 1.20 × calculated volume
Evaporator—estimated volume = 1.20 × calculated volume
Economizer—estimated volume = 1.20 × calculated volume
Condenser—estimated volume = 1.20 × calculated volume
Recuperator—estimated volume = 1.05 × calculated volume
Intercooler—estimated volume = 1.05 × calculated volume

4.5 Predicted Performance of Small Modular NACC systems

To assess the performance of small modular NACC systems two technology levels were considered. A near-term system was represented by a sodium reactor with an input temperature of 783 K to the first turbine and an advanced system was represented by a molten salt reactor with an input temperature of 948 K to the first turbine. Both systems were designed to produce 50 MW(e). Up to four reheat cycles, or five primary heat exchangers and turbines, were considered for both systems. The input and exit temperatures for all turbines were the same. The performance characteristics for the near-term sodium systems are given in Table 4.1.

Note that the efficiencies are not much better than current LWR systems. The limitation on temperature is severe when compared to a GTCC. Even though the efficiencies are not any better than a current LWR system, the heat dumps to environmental water (15.8–21.5 MW) are significantly less than a 35% efficient LWR that would have to dump 92.8 MW of heat to produce 50 MW(e).

The performance characteristics for an advanced Air-Brayton system based on a molten salt coolant are given in Table 4.2. Now the efficiencies are significantly better but still not up to the gas turbine systems. However,

Table 4.1 Performance characteristics of a sodium air-Brayton combined cycle system.

Characteristic	2 turbines	3 turbines	4 turbines	5 turbines
Electrical power, MW(e)	50	50	50	50
Efficiency	30.44%	34.10%	35.92%	37.14%
Thermal power, MW(t)	164.3	146.6	139.2	134.6
Critical power ratio (CPR)	4.957	7.541	9.703	10.28
T(Turbine inlet), K	783	783	783	783
T(Turbine exit), K	648	668	683	703
Mass flow rate air, kg/s	333.2	272.9	245.2	227.4
Mass flow rate water, kg/s	7.6	8.3	8.9	9.6
Brayton power, MW(e)	39.2	37.9	36.7	35.2
Rankine power, MW(e)	10.8	12.1	13.3	14.8
Water heat dump, MW(t)	15.8	17.7	19.6	21.5
Reactor size, cu m	236	218	210	206
HRSG, cu m	39	35	35	36
Recuperator, cu m	0	0	0	0
Brayton system, cu m	10.9	10.8	11.6	12.9
Rankine system, cu m	46	43	44	46
System volume, cu m	304	282	276	275

Table 4.2 Performance characteristics of a molten salt air-Brayton combined cycle system.

Characteristic	2 turbines	3 turbines	4 turbines	5 turbines
Electrical power, MW(e)	50	50	50	50
Efficiency, %	43.24	45.36	46.72	47.48
Thermal power, MW(t)	115.6	110.2	107.0	105.3
CPR	9.301	11.917	14.478	16.928
T(turbine inlet), K	948	948	948	948
T(turbine exit), K	728	783	813	833
Mass flow rate air, kg/s	172.6	147.9	135	127.9
Mass flow rate water, kg/s	9.5	10.5	10.7	10.8
Brayton power, MW(e)	35.3	32.4	31.3	30.7
Rankine power, MW(e)	14.7	17.6	18.7	19.3
Water heat dump, MW(t)	22.4	25.3	26	26.4
Reactor size, cu m	186	181	178	176
HRSG, cu m	31	34	35	36
Recuperator, cu m	0	0	0	0
Brayton system, cu m	66	7.3	8.8	19.5
Rankine system, cu m	41	45	47	48
System volume, cu m	243	243	243	245

it is probably not a good idea to compare a small modular nuclear reactor against a gas turbine system on any performance measure other than the cost of electricity.

Efficiency is everything for a gas turbine system, but it is not that significant for a nuclear system as the cost of fuel is a much smaller fraction of the cost of electricity for a nuclear system. It is also worth pointing out that these are the highest pressure systems that will be addressed here. The peak pressure is about 17 atmospheres, or about 1.7 MPa. This is significantly less than current LWR system pressures and quite a bit less than the 7 MPa+ being proposed for supercritical CO_2 systems.

The water heat dumps for these systems are larger than for the sodium systems because a larger fraction of the energy is derived from the bottoming cycle.

The efficiency of the near-term sodium CCs can be improved by adding a recuperator after the HSRG. This is a relatively small recuperator compared to an NARC system type, but it does boost the efficiency. Its performance characteristics are given in Table 4.3.

Note that the efficiency is not a strong function of the number of turbines or reheat cycles and the two turbine system is competitive with the three-, four-, and five-turbine systems. The system pressures are also lower, as

Table 4.3 Performance characteristics of a sodium air-Brayton combined cycle system with an added recuperator.

Characteristic	2 turbines	3 turbines	4 turbines	5 turbines
Electrical power, MW(e)	50	50	50	50
Efficiency, %	40.70	40.98	41.00	40.98
Thermal power, MW(t)	122.9	122.0	122.0	122.0
CPR	2.598	2.556	2.609	2.788
T(turbine inlet), K	783	783	783	783
T(turbine exit), K	713	730	743	748
Mass flow rate air, kg/s	351.7	309.2	293	277
Mass flow rate water, kg/s	16.4	16.1	16.5	16
Brayton power, MW(e)	24.3	24	23	23.5
Rankine power, MW(e)	25.7	26	27	26.5
Water heat dump, MW(t)	29.8	36.6	37.4	36.5
Reactor size, cu m	194	194	194	194
HRSG, cu m	59	57	57	55
Recuperator, cu m	74	65	62	58
Brayton system, cu m	82	76	75	73
Rankine system, cu m	77	75	76	73
System volume, cu m	374	364	364	360

is characteristic of any recuperated system. Once again, the heat dumps are a little larger because the bottoming cycle is doing more work. The advanced molten salt system could not accept a recuperator because the compressor exit temperatures were greater than the final turbine exit temperatures.

4.6 Performance Variation of Small Modular NACC Systems

Given the performance described in Tables 4.1 to 4.3 it is worth considering the sensitivity of these systems to some of the assumptions made in the analysis. As it gets rather messy if one tries to assess the sensitivity of three different systems with four-turbine configurations each, a three-turbine system will be chosen to assess sensitivity and the recuperator will be included in the near-term system.

The first most obvious sensitivity would be to the temperature of the air entering the turbines. This has been addressed already by considering two baseline temperatures and linear interpolation between the two should be adequate. The remaining two most sensitive parameters driving the performances of these NACC systems are the steam pressure in the bottoming cycle and the effectiveness of the recuperator in the sodium system. The sensitivity to recuperator effectiveness will be addressed when discussing

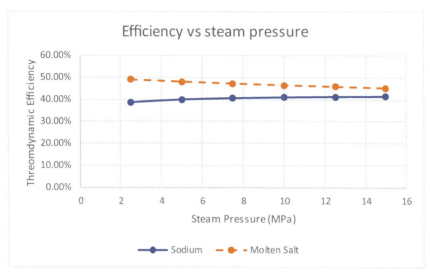

Fig. 4.6 Thermodynamic efficiency versus steam pressure for air-Brayton combined cycles.

NARC systems. The sensitivity of thermodynamic efficiency to the steam pressure for both systems is presented in Fig. 4.6.

As the sodium system is recuperated and the molten salt system is not recuperated, the effects of steam pressure on each seem to be in the opposite direction. However, it is not a big effect. The effect of steam pressure on the HRSG size is presented in Fig. 4.7.

Once again it would seem that the effect of recuperation tends to moderate the increase in size of the HRSG with steam pressure, though this time

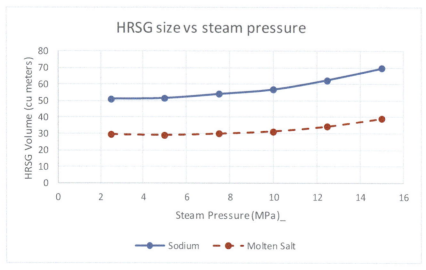

Fig. 4.7 Variation of HRSG size versus steam pressure for Air-Brayton combined cycles.

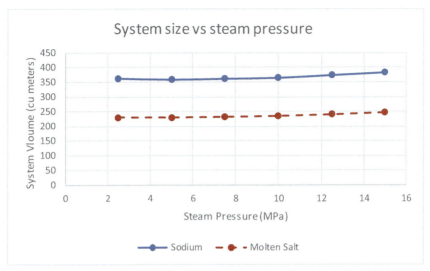

Fig. 4.8 System size versus steam pressure for Air-Brayton combined cycles.

the variation is in the same direction increasing with increased pressure. The effect on the overall system size is presented in Fig. 4.8.

The size of the HRSG is not large enough to influence the overall system size greatly for different steam pressures. It and the recuperator are smaller than the estimate for the reactor and primary heat transport system.

As the baseline steam turbines were assumed to be quite efficient, it was considered useful to see the effect of this assumption on the overall system. The sensitivity of the overall system thermodynamic efficiency is presented in Fig. 4.9.

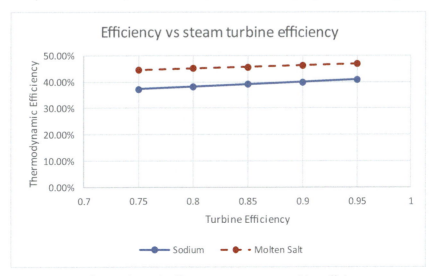

Fig. 4.9 System thermodynamic efficiency versus steam turbine efficiency.

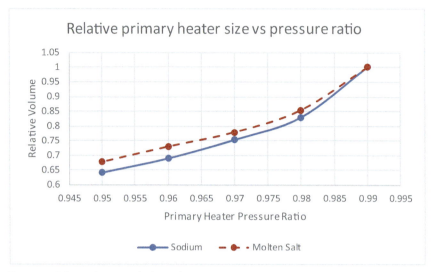

Fig. 4.10 Primary heater relative volume versus pressure ratio.

The system efficiency decreases 0.18% for every 1% decrease in the steam turbine efficiency for the near-term sodium system, and it decreases 0.12% for every 1% decrease in the steam turbine efficiency for the advanced molten salt system. Thus, the steam turbine efficiency is not a major determinant of overall efficiency.

Next consider the effect of the primary heater pressure drop or pressure ratio on heater size and system size. The size of the heaters decreases by 35–37% if the allowed pressure ratio goes from 0.99 to 0.95 as described in Fig. 4.10. It has a larger impact on the overall system size than might be expected as the pressure losses in the heaters affect the sizes of everything down stream. It is a significantly bigger effect for the near-term sodium system than for the advanced molten salt system. The magnitude of this effect is plotted in Fig. 4.11.

The primary heater pressure ratio affects the system efficiency in a linear manner with a decrease in system efficiency of 0.73% for every 1% decrease in the pressure ratio for the near-term system, and a 0.35% decrease for every 1% decrease in pressure ratio for the advanced system.

Next consider the HSRG. This consists of the economizer, evaporator, and the superheaters. It is simpler to just lump all of the effects together as all are heat exchangers and the pressure drop and effectiveness are the parameters of interest. In Figs. 4.12 and 4.13 all of these heat exchanger parameters are varied the same.

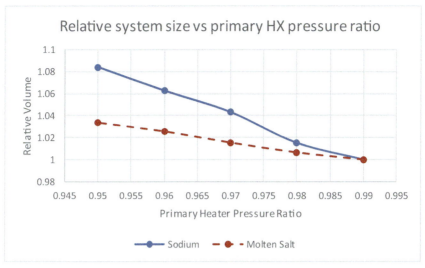

Fig. 4.11 Relative system volume versus primary heater pressure ratio.

Fig. 4.12 gives the variation in the relative volume for the HRSG as a function of all of the pressure ratios.

Fig. 4.13 gives the relative variation in volume for the HRSG as a function of the effectiveness of all of its components.

The relative variation in the size of the HRSG is the same for both near-term and advanced systems resulting in overlapping curves in Figs. 4.12 and 4.13.

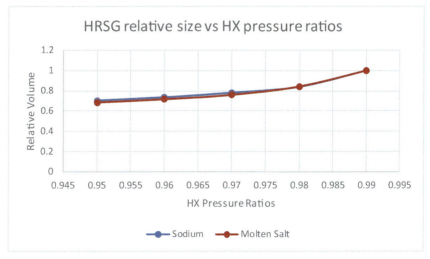

Fig. 4.12 HRSG relative size versus the pressure ratios for all of its components.

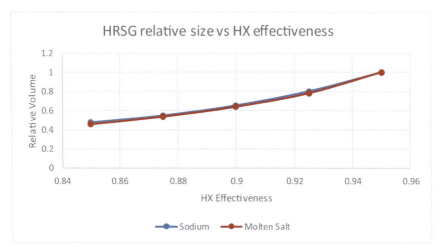

Fig. 4.13 HRSG relative volume as a function of its components effectiveness.

The pressure drop through the HRSG components affects the system efficiency in exactly the same manner as the pressure drop through the primary heaters—0.73% per 1% for sodium and 0.34% per 1% for molten salt.

The system efficiency is affected less by the effectiveness of the HRSG than the pressure drops through it. The loss in system efficiency for a 1% change in the effectiveness is 0.31% for the near-term system and 0.30% for the advanced system.

This concludes the sensitivity analysis for NACC systems. The sensitivity to recuperator performance will be addressed in the next section when NARC systems are discussed.

4.7 Predicted Performance for Small Modular NARC Systems

Small modular NARC systems are interesting for a number of reasons. They can achieve higher thermodynamic efficiencies than NACC systems, with an advanced system with intercooler having a predicted efficiency of greater than 50%. They also can be built without a water heat dump for waste heat, making them locatable anywhere on the planet.

Consider a typical near-term NARC system that produces 50 MW(e) in Table 4.4. Note that efficiency drops off slightly with an increase in the number of turbines, though it is about 1% better than the best recuperated NACC system, and 4% better than the best basic NACC system. The pressures are very comparable to the recuperated near-term NACC system, and the relative volumes are larger than the basic NACC system but smaller

Table 4.4 Performance characteristics of a sodium air-Brayton recuperated system.

Characteristic	2 turbines	3 turbines	4 turbines	5 turbines
Electrical power, MW(e)	50	50	50	50
Efficiency, %	42.61	42.58	42.30	41.91
Thermal power, MW(t)	117.3	117.4	118.2	119.3
CPR	2.497	2.67	2.795	2.935
T(turbine inlet), K	783	783	783	783
T(turbine exit), K	703	725	738	745
Mass flow rate air, kg/s	602.9	558.6	538	521.2
Mass flow rate water, kg/s	0	0	0	0
Brayton power, MW(e)	50	50	50	50
Rankine power, MW(e)	0	0	0	0
Water heat dump, MW(t)	0	0	0	0
Reactor size, cu m	188	188	189	190
HRSG, cu m	0	0	0	0
Recuperator cu m	107	97	93	89
Brayton system, cu m	145	118	117	117
Rankine system, cu,	0	0	0	0
System volume	321	315	314	315

than the recuperated NACC system. Mass flow rates are almost double those of the NACC systems.

The performance characteristics of an advanced NARC system are given in Table 4.5.

Table 4.5 Performance characteristics of a molten salt air-Brayton recuperated system.

Characteristic	2 turbines	3 turbines	4 turbines	5 turbines
Electrical power, MW(e)	50	50	50	50
Efficiency, %	49.55	49.40	49.24	48.99
Thermal power MW(t)	100.9	101.2	101.5	102.1
CPR	2.847	3.077	3.345	3.365
T(turbine inlet), K	948	948	948	948
T(turbine exit), K	855	868	883	898
Mass flow rate air, kg/s	377.9	348.7	326.2	326.1
Mass flow rate water, kg/s	0	0	0	0
Brayton power, MW(e)	50	50	50	50
Rankine power, MW(e)	0	0	0	0
Water heat dump, MW(t)	0	0	0	0
Reactor size, cu m	171	172	172	173
HRSG, cu m	0	0	0	0
Recuperator, cu m	62	56	51	51
Brayton system, cu m	73	70	67	70
Rankine system, cu m	0	0	0	0
System volume, cu m	249	246	244	248

A comparison of the advanced NARC system with the advanced NACC system shows many of the same changes that the near-term comparison showed. However, the advanced NACC system could not use a recuperator, so the NARC system operates at a much lower pressure and gets about a 2% better thermodynamic efficiency.

4.8 Performance Variation of Small Modular NARC Systems

There are only two components that drive the sensitivities of NARC systems—the primary heat exchanger and the recuperator. Both are sensitive to the pressure drop and the effectiveness of the heat exchange process. Start with the primary heat exchanger. The relative volume as a function of the primary heat exchanger pressure ratio is described in Fig. 4.14.

As expected, the effect on relative volume is the same for both the near-term and advanced systems, however the effect of a 5% reduction in pressure ratio produces about a 5% greater savings in relative volume for the NARC system over the NACC system.

The change in primary heater pressure ratio has a greater effect on the thermodynamic efficiency for the NARC systems than it did for the NACC systems. In this case the system thermodynamic efficiency for the near-term system decreases 1.3% for every 1.0% decrease in the primary heater pressure ratio. For the advanced system this decrease in thermodynamic efficiency is 1% for each 1% decrease in primary heater pressure ratio. This is the largest change in system efficiency observed when considering the variation in component properties.

The change in system relative volume as the primary heater pressure ratio is varied is described in Fig. 4.15. As before for the NACC systems,

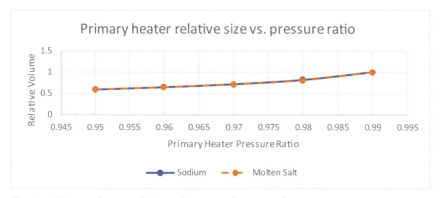

Fig. 4.14 Primary heater relative volume as a function of its pressure ratio.

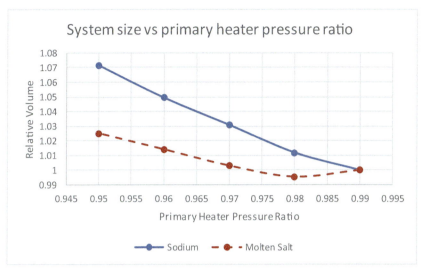

Fig. 4.15 System relative volume as a function of primary heater pressure ratio.

the variation in primary heat exchanger pressure ratio has a much bigger impact on overall system volume increases for the near-term sodium system than it does for the advanced molten salt system.

Fig. 4.16 describes the effect of primary heat exchanger effectiveness on the size of the heat exchangers. Reducing the effectiveness by 20% reduces the size of the heat exchangers by 60%. This is essentially the same reduction achieved for the NACC systems. Contrary to the effect of the primary

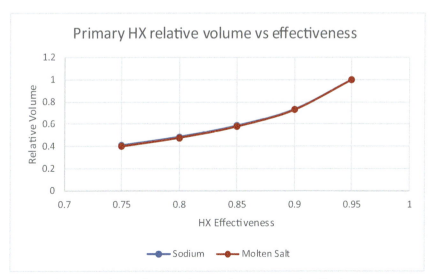

Fig. 4.16 Relative volume of primary heat exchanger versus its effectiveness.

heat exchanger pressure ratio, the change in the effectiveness of the primary heat exchanger has a negligible effect on the system volume. The primary heat exchanger effectiveness affects only its size and does not affect the size of other system components. Once again, the two curves for the near-term and advanced systems are identical.

Given that the temperatures into the turbines were specified for this analysis, it is impossible to determine the effect of the primary heat exchanger effectiveness on the system thermodynamic efficiency.

The effects of the recuperator pressure ratio and effectiveness on system size and efficiency for an NARC system are significant. First consider the effect of the pressure ratio. The effect of pressure ratio on recuperator size is described in Fig. 4.17.

Unlike the primary heat exchangers, the recuperator variation in pressure ratio has very little effect on the system size, less than 1% for the 5% variation in pressure ratio.

The thermodynamic efficiency varies quite linearly with recuperator pressure ratio. The system efficiency decreases 0.9% for every 1% decrease in the pressure ratio for the near-term sodium system and 0.67% for every 1% for the advanced molten salt system.

The recuperator relative volume as a function of its effectiveness is described in Fig. 4.18.

There is a dramatic drop in size by decreasing the required effectiveness by 5%. The 0.9 effective recuperator is only 20% the size of the 0.95

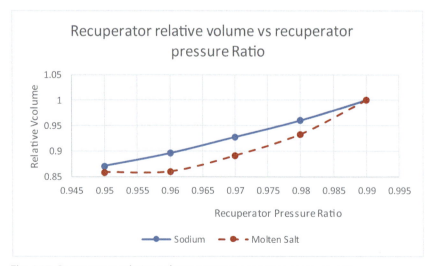

Fig. 4.17 Recuperator relative volume versus recuperator pressure ratio.

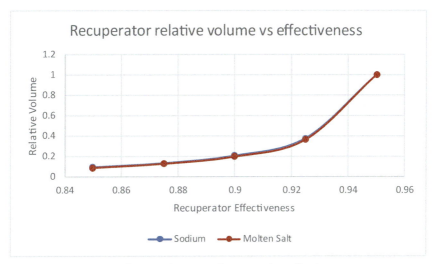

Fig. 4.18 Recuperator relative volume as a function of its effectiveness.

effective recuperator. Another 10% decrease in size can be achieved by lowering the effectiveness another 5%, but beyond that the drop in size levels out and is not as dramatic. Note once again that the curves for the near-term and advanced systems are coincident.

The effect on the overall NARC system by adjusting the recuperator effectiveness is described in Fig. 4.19.

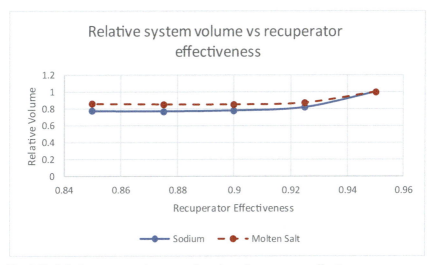

Fig. 4.19 Relative system volume as a function of recuperator effectiveness.

For these cases, the recuperators are not a large fraction of the total system volume. The design chosen for the recuperators is very aggressive, and as a recuperator becomes a larger part of the volume of the system, the impact of its change in the system volume will be much more significant.

The effect of the recuperator effectiveness on the system thermodynamic efficiency is 0.77% per 1% for the near-term sodium system and 0.89% per 1% for the advanced molten salt system.

4.9 Predicted Performance for a Small Modular Intercooled NARC System

All Air-Brayton systems lose efficiency as more work is required to compress the air. As the air is compressed, it heats up and it takes increasingly more work to compress it. One solution to overcoming this limitation is to split the compressor in two and cool the air after it leaves the first half of the compressor. This results in the second half of the compressor working on cooler air and delivering cooler air to the primary heat exchangers. As the air must now be heated more in the primary heat exchangers, the efficiency goes down. But if the system is recuperated, the recuperator can usually put back in the lost heat. In other words, intercooling is a good idea to improve efficiency, but to achieve the desired efficiency improvement, the system must have a recuperator.

The system characteristics for a near-term intercooled NARC system are presented in Table 4.6. Once again, the thermodynamic efficiency does

Table 4.6 Performance characteristics of a sodium air-Brayton intercooled NARC system.

Characteristic	2 turbines	3 turbines	4 turbines	5 turbines
Electrical power, MW(e)	50	50	50	50
Efficiency	45.11%	45.48%	45.51%	45.38%
Thermal power, MW(t)	110.8	109.9	109.9	110.2
CPR	3.251	3.595	4.119	4.429
T(turbine inlet), K	783	783	783	783
T(turbine exit), K	680	705	720	730
Mass flow rate air, kg/s	454.6	398.1	370.1	353.3
Mass flow rate water, kg/s	0	0	0	0
Brayton power, MW(e)	50	50	50	50
Rankine power, MW(e)	0	0	0	0
Water heat dump, MW(t)	24.1	24	24.1	24.3
Reactor size, cu m	181	181	180	181
HRSG, cu m	0	0	0	0
Recuperator, cu m	81	69	63	60
Brayton system, cu m	127	111	105	102
Rankine system, cu,	0	0	0	0
System volume	316	299	292	290

Table 4.7 Performance characteristics of a molten salt air-Brayton intercooled system.

Characteristic	2 turbines	3 turbines	4 turbines	5 turbines
Electrical power, MW(e)	50	50	50	50
Efficiency, %	52.01	52.36	52.46	52.43
Thermal power, MW(t)	96.1	95.5	95.3	95.4
CPR	3.949	4.594	5.138	5.381
T(turbine inlet), K	948	948	948	948
T(turbine exit), K	808	843	863	878
Mass flow rate air, kg/s	282.6	249.5	230.8	223.9
Mass flow rate water, kg/s	0	0	0	0
Brayton power, MW(e)	50	50	50	50
Rankine power, MW(e)	0	0	0	0
Water heat dump, MW(t)	17.9	17.8	17.9	17.9
Reactor size, cu m	167	166	166	166
HRSG, cu m	0	0	0	0
Recuperator, cu m	46	38	36	35
Brayton system, cu m	75	67	64	64
Rankine system, cu	0	0	0	0
System volume	246	237	233	234

not vary significantly based on the number of reheat cycles or turbines involved. The efficiencies are about 3% higher and the pressures have increased by 0.75 –1.5 atmospheres. The system sizes are very comparable to the nonintercooled NARC system with the recuperator shrinking to make room for the intercooler.

An advanced molten salt intercooled NARC system is characterized in Table 4.7. This is the system that finally achieves an efficiency greater than 50%.

And once again the system efficiency is not greatly dependent on the number of reheat cycles or turbines. The efficiencies are up by ~2.5%. The pressures are up by 1–2 atmospheres. The mass flows are about two-thirds the nonintercooled advanced NARC system. And the system volumes are very comparable, with the intercooled systems slightly smaller.

4.10 Performance Variation of Small Modular Intercooled NARC Systems

As the recuperator and primary heat exchanger have been addressed for the NARC system before, only the sensitivities to the intercooler pressure ratio and effectiveness will be addressed here. The intercooler pressure ratio has only a moderate effect on its size as described in Fig. 4.20. Note also that the effect is the same for both the near-term and advanced systems.

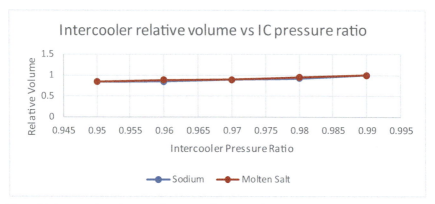

Fig. 4.20 Intercooler relative volume as a function of its pressure ratio.

The system thermodynamic efficiencies once again look linear as a function of intercooler pressure ratio with the near-term system decreasing 0.31% for every 1% decrease in the pressure ratio. The advanced system efficiency decreases 0.24% for every 1% decrease in the intercooler pressure ratio.

Somewhat similar to the recuperator, the intercooler volume decreases rather significantly as a function of its effectiveness. The relationship describing this sensitivity is described in Fig. 4.21. In dropping 20% in effectiveness, the size of the intercooler decreases by 81%. However, this does not have a significant effect on the overall system size as the

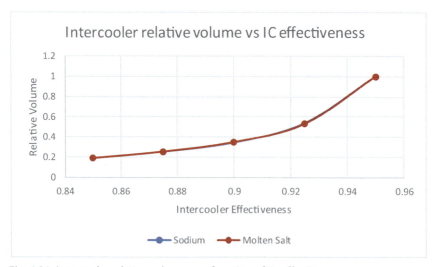

Fig. 4.21 Intercooler relative volume as a function of its effectiveness.

recuperator compensates somewhat. The overall system decreases in size by less than 1.5% for both systems as the intercooler size drops from its nominal value to 19% of that value.

The thermodynamic efficiency of an intercooled NARC system is least sensitive to the effectiveness of the intercooler. For every 1% decrease in the effectiveness of the near-term intercooler, the thermodynamic efficiency only decreases 0.084%. For the advanced system, a 1% decrease in the effectiveness only generates 0.083% decrease in the thermodynamic efficiency of the whole system.

4.11 Conclusions

The performance of NACC and NARC systems have been estimated for a near-term sodium-cooled system and an advanced molten salt cooled system. Both systems achieve reasonable thermodynamic efficiencies significantly better than current LWR systems. However, efficiencies may not be the driver in the development of advanced SMRs. The sensitivities of efficiency and size have been considered for all of the major components of NACC and NARC systems. Nuclear Air-Brayton systems are very similar to gas turbine systems and can adapt much of the technology from these systems.

The industrial base for gas turbines is very large. NACC and NARC systems require significantly less environmental water to absorb their waste heat, with the simple recuperated system requiring zero. As the air working fluid of the Nuclear Air-Brayton systems is not consumed, it may be reheated and expanded several times. This adds a great deal of flexibility to NACC and NARC systems.

Additional advantages are discussed in Chapter 5 and Appendix A of this book as well in Zohuri and McDaniel book [11].

References

[1] American Nuclear Society, Nuclear News, 21st Reference Issue, March 2019.
[2] J Buongiorno, M. Corradini, J. Parsons, D. PettiCo-Chairs, The Future of Nuclear Energy in a Carbon-Constrained World, an Interdisciplinary Study, Massachusetts Institute of Technology, 2018.
[3] B. Zohuri, P. McDaniel, Combined Cycle Driven Efficiency for Next Generation Nuclear Power Plants, An Innovative Design Approach, Second ed, Springer nature, Switzerland, 2018.
[4] M.M. Wakil, Powerplant Technology, McGraw-Hill International Editions, New York, 1984.

[5] V. Dostal, M.J. Driscoll, P.A. Hejzlar, A Supercritical Carbon Dioxide Cycle for Next Generation Nuclear Reactors, Tech Rep MIT-ANP-TR-100, Massachusetts Institute of Technology, Cambridge, MA, 2004.
[6] A.E. Waltar, D.R. Todd, P.V. Tsvetkov, Fast Spectrum Reactors, Springer Science, NY, 2012.
[7] D.G. Wilson, T. Korakianitis, The Design of High-Efficiency Turbomachinery and Gas Turbines, 2nd Ed, Prentice-Hall, Inc., Upper Saddle River, NJ, 1998.
[8] S.A. Korpela, Principles of Turbomachinery, John Wiley & Sons, Inc, Hoboken, NJ, 2011.
[9] T. Blumberg, M. Assar, T. Morosuk, G. Tsatsaronis, Comparative exergoeconomic Evaluation of The Latest Generation of Combined-Cycle Power Plants, Energy Convers. Manag. 153 (2017) 616–626.
[10] W.M. Kays, A.L. London, Compact Heat Exchangers, 3rd Ed, Krieger Publishing Company, Malabar, FL, 1998.
[11] B. Zohuri, P.J. McDaniel, Advanced Smaller Modular Reactors: An Innovative Approach to Nuclear Power, First ed, Springer Publishing Company, NY, 2019.

CHAPTER 5

Advanced Nuclear Open Air-Brayton Cycles for Highly Efficient Power Conversion

5.1 Introduction

The nuclear air–Brayton combined cycle (NACC) and the nuclear air–Brayton recuperated cycle (NARC) are excellent potential power conversion systems for the next-generation nuclear reactors. They provide a high thermal conversion efficiency and the possibility of a significantly smaller component footprint than current light water systems. Both characteristics should produce significant economic advantages. They take advantage of the large investment by the aerospace and gas turbine power conversion industries in air compressors and turbines. The NACC system significantly reduces the cooling water waste heat requirement and the NARC system eliminates it. This would allow their economic deployment to a much larger fraction of the earth's surface.

The study undertaken here and presented below focused on a 25-MW (electric) power conversion module. It is anticipated that one to four modules could be used with a small modular reactor (SMR), giving a range of power from 25 to 100 MW(electric). This represents the bottom of the power range for SMRs. This is also probably the smallest module that might be produced economically. It also falls in the range of power level that might take the most advantage of the large base of components and parts in the aerospace industry. The size of the compressor and turbine components for this size of module are comparable to those on the engines of a Boeing 737. It is also representative of the size that might be developed to demonstrate the performance of one of these systems.

Scaling laws are presented to scale the volume of components to higher powers, with the implication that the fixed plant cost would scale comparably [23].

5.2 Background

The major growth in the electricity production industry in the last 30 years has centered on the expansion of natural gas power plants based on gas turbine cycles. The most popular extension of the simple Brayton gas turbine has been the combined cycle power plant with the Air-Brayton cycle serves as the topping cycle and the steam-Rankine cycle serves as the bottoming cycle. The air-Brayton cycle is an open air cycle and the steam-Rankine cycle is a closed cycle. The air-Brayton cycle for a natural gas–driven power plant must be an open cycle, where the air is drawn in from the environment and exhausted with the products of combustion to the environment. The hot exhaust from the air-Brayton cycle passes through a heat recovery steam generator (HRSG) prior to exhausting to the environment in a combined cycle [1,2]. The HRSG serves the same purpose as a boiler in the conventional steam-Rankine cycle.

In 2007, gas turbine with combined cycle plants had a total capacity of 800 GW and represented 20% of the installed capacity worldwide. They have far exceeded the installed capacity of nuclear plants, though in the late 1990s they had less than 5% of the installed capacity worldwide (Ref. 3, Table 2.1, p. 7). There are a number of reasons for this. First the natural gas is abundant and cheap. Second combined cycle plants achieve the greatest efficiency of any thermal plant (Ref. 3, Table 3.2, p. 21). And third, they require the least amount of waste heat cooling water of any thermal plant.

A typical gas turbine plant consists of a compressor, combustion chamber, turbine, and an electrical generator. The working fluid is air [4-6]. The gas turbine exhaust is ducted to the HRSG.

The HRSG serves as the boiler in a typical closed-cycle steam plant. A steam plant consists of a steam turbine, a condenser, a water pump, an evaporator (boiler), and an electrical generator. The working fluid is water [7]. In a combined cycle plant, the gas turbine and steam turbine can be on the same shaft to eliminate the need for one of the electrical generators. However, two-shaft two-generator systems provide a great deal of more flexibility at a slightly higher cost [8].

In addition to the closed loop for the steam, an open-loop circulating water system is required to extract the waste heat from the condenser. The waste heat extracted by this circulating water system is significantly less than that in a conventional thermal plant as the open air-Brayton cycle exhausts its waste heat directly to the air.

The layout for a typical combined cycle power plant is given in Fig. 5.1.

The approximate efficiency can be calculated for a combined cycle power plant by the following simple argument (Ref. 2, p. 38):

$$\text{Brayton cycle efficiency} = \frac{W_B}{Q_{in}} = \eta_B \quad (5.1)$$

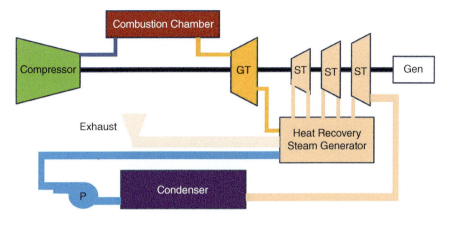

Fig. 5.1 Typical gas turbine combined cycle power plant.

$$\text{Heat to Rankine cycle} = Q_R = (1-\eta_B)Q_{in} \quad (5.2)$$

$$\text{Rankine cycle efficiency} = \frac{W_R}{Q_R} = \eta_R \quad (5.3)$$

$$\text{Overall efficiency} = \frac{W_B + W_R}{Q_{in}} = \eta_T = \frac{\eta_B Q_{in} + \eta_R Q_R}{Q_{in}}$$

$$= \frac{\eta_B Q_{in} + \eta_R (1-\eta_B)Q_{in}}{Q_{in}} \quad (5.4)$$

$$= \eta_B + \eta_R - \eta_B \eta_R$$

$$\eta_T = \eta_B + \eta_R - \eta_B \eta_R$$

This efficiency has to be corrected for pressure losses and assumes that all of the heat in the Brayton exhaust is used in the HRSG. For a combustion gas turbine, this is not usually possible if condensation of the water in the exhaust products is to be avoided. The detailed models developed in this effort give a more accurate answer.

For a nuclear system to take advantage of combined cycle technology, there are a number of changes to the plant components that have to be made. The most significant, of course, is that the combustion chamber has to be replaced by a heat exchanger in which the working fluid from the nuclear reactor secondary loop is used to heat the air. The normal Brayton cycle is an internal combustion one. The working fluid is heated by the combustion of the fuel-air mixture in the combustion chamber. Four-fifths of the air (nitrogen) does not participate in the combustion. The walls of the combustion chamber can be cooled and peak temperatures in the working fluid can be significantly above the temperature that the walls of the chamber can tolerate for any length of time [6,9,10].

For a nuclear reactor system, the heat transfer is in the opposite direction. All reactor components and fluids in the primary and secondary loops must be at a higher temperature than the peak temperature of the gas exiting the heat exchanger. This severely restricts the peak temperature that can be achieved for the air entering the turbine. However, all is not lost.

In a typical combustion system, there are pressure losses approaching 5% of the total pressure needed to complete the combustion process (Ref. 9, Table 6.2, p. 363 and Ref. 10, Fig. 6.20, p. 220). Heat exchangers can be built with significantly lower pressure drops than 5% of the entering pressure, possibly approaching 1% (Ref. 11). The most straightforward method to overcome this severe temperature limitation is to borrow a technique from steam power plants and implement multiple reheat cycles. That is, the first heat exchanger heats the air to its peak temperature. Then the air is expanded through the first turbine. The air is then reheated to the same peak temperature and expanded through the second turbine. Based on the relative pressure losses that appear possible, up to five turbines might be considered. All five turbines would drive the same compressor [12].

Multiple compressors on concentric shafts driven by different sets of turbines might be possible, but that has not been considered here. Multiple reheat cycles allow more heat to be put into the working fluid at a higher temperature. This improves the efficiency of the overall cycle. Interestingly enough, this does not improve the efficiency of the Brayton cycle, but because the exit temperature from the last turbine is higher, it does improve the efficiency of the Rankine part of the cycle with a net gain for the overall cycle. For this to work, the reactor coolant temperature must reach temperatures significantly higher than current light water reactor (LWR) temperatures. Even sodium-cooled reactor exit temperatures in the 550 °C range are not quite high enough to get an NACC to compete with the efficiency of a pure steam-Rankine cycle. But when the coolant exit temperatures reach the 650–700 °C range, the combined cycle systems, with multiple turbines, surpass the performance of steam-Rankine systems. Therefore, the analysis that follows is targeted at a fluoride salt reactor [13] and a lead coolant reactor [14]. A pressurized sodium reactor that could reach these temperatures is another possibility, but no one is proposing such a system at this time. It could also apply to a high-temperature gas reactor, but the heat exchangers would be quite different. Gas-to-gas heat exchangers have not been considered as the primary heat exchangers at this point but will be addressed in the recuperated systems.

Liquid metal and molten salt heat exchangers were developed and tested successfully [11,15-17] during the Aircraft Nuclear Propulsion program in the late 1950s. They were conventional tube and plate exchangers and were tested for over 1000 h at temperatures up to 1100 K. The largest size tested transferred 55 MW of heat in a package of approximately 1.2 m^3. The heat transfer area on the air side had a surface area per unit volume of 1200 m^2/m^3. Certainly, some development will be needed to bring this technology up to modern standards, but the tasks involved do not appear insurmountable.

A number of additional heat exchangers were designed in this work in an attempt to estimate sizes of components and validate that pressure drop criteria could be met. The heat from the exhaust of the air-Brayton cycle transfers heat to vaporize the steam in the steam-Rankine cycle in an HRSG of a fairly conventional design. This includes air-to-steam superheaters as well as an economizer and evaporator section. A condenser of conventional design is included. For this work, all heat exchangers were considered to be counterflow designs [24].

The heat exchanger design procedures and experimental data were taken from the text by Kays and London [18]. All of the data presented in this text were developed from steam-to-air heat exchangers and should be particularly applicable to the types of heat exchangers used in this analysis. The only ones not using these two fluids are the molten salt (or liquid metal)-to-air heat exchangers referenced above. Thus, this is a real basis for all of the design calculations used to estimate power conversion system performance and sizing.

One of the significant advantages of the combined cycle power system over current LWR power systems is its reduced requirement for circulating water in the waste heat rejection loop for the steam-Rankine cycle. The typical combined cycle plant considered here produces approximately 50% of its power from the steam-Rankine cycle and 50% of its power from the air-Brayton cycle.

This automatically reduces the cooling water requirement by half. In addition, the combined cycle plant achieves 45% efficiency so that only 55% of the heat generated has to be released as waste. A typical 25-MW(electric) system will only need to get rid of 6.9 MW of heat via a circulating water system. A current LWR plant generating 25 MW (electric) at an efficiency of 33% would need to dump 16.8 MW of heat. This represents a major saving in circulating water requirements.

As the combined cycle reduces the circulating water requirement so significantly, the natural question arose as to whether it could be eliminated completely. With a recuperated air-Brayton cycle, it can be. All of the waste heat can be rejected directly to the atmosphere. At first it was thought that the efficiency of a multiturbine recuperated cycle could not compete with a combined cycle plant. However, after performing the detailed analysis, the efficiencies of a recuperated cycle system come within 1% or 2% of the predicted combined cycle efficiencies. This would seem to be a minor penalty to pay for being free of a circulating water requirement. However, achieving these high efficiencies requires a very effective recuperator, which can become quite large.

5.3 Combined Cycle Feature

Both the air-Brayton cycle and the steam-Rankine cycle, in the combination of the gas turbine, complement each other to form efficient combined cycles. The most commonly used working fluids for combined cycles are air and steam. Other working fluids (organic fluids, potassium vapor, mercury vapor, and others) have been applied on a limited scale.

Combined cycle systems that utilize air and steam as working fluids have achieved widespread commercial application due to [7].
1. high thermal efficiency through application of two complementary thermodynamic cycles
2. heat rejection from the Brayton cycle (gas turbine) at a temperature that can be utilized in a simple and efficient manner
3. readily available, inexpensive, and nontoxic working fluids (water and air).

5.4 Typical Cycles

Figs. 5.2 and 5.3 provide a schematic of a four-turbine combined cycle system and its thermodynamic cycles on a temperature-entropy plot. This turns out to be the near-optimum combined cycle system based on the analysis presented below.

Figs. 5.4 and 5.5 provide a schematic layout and a temperature-entropy diagram for a three-turbine recuperated system. This turns out to be the optimum recuperated system.

Advanced nuclear open air-brayton cycles for highly efficient power conversion   177

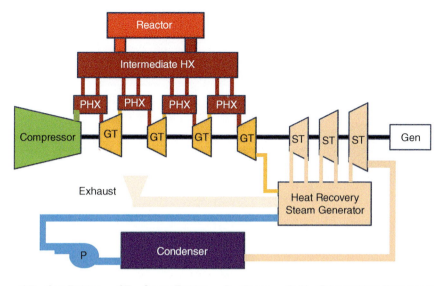

GT = Gas Turbine   ST = Steam Turbine   P = Pump   PHX = Primary Heat Exchanger

Fig. 5.2 Layout for four-turbine combined cycle.

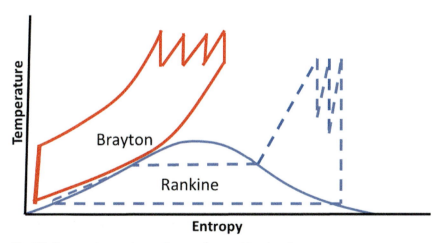

Fig. 5.3 Temperature-entropy diagram for combined cycle.

5.5 Analysis Methodology

The approach taken for this effort was to model the thermodynamics of the components making up the power conversion systems as real components with nonideal efficiencies [25]. Pressure drops are included for every component except the connected piping. The compressor design is modeled

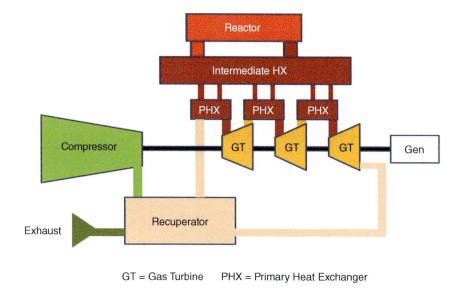

Fig. 5.4 Recuperated system layout schematic.

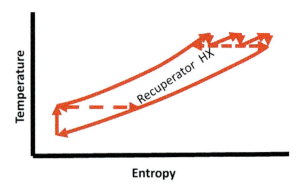

Fig. 5.5 Recuperated system temperature-entropy diagram.

with a small-stage polytropic efficiency to take into account state-of-the-art designs (Ref. 5, pp. 79–80 and Ref. 9, Table 6.2, p. 363). The gas turbines are likewise modeled with a polytropic efficiency. The steam turbines are modeled with a simple overall thermal efficiency. Pressure drops in each of the heat exchangers are included. The input to the analysis specifies the pressure drops, and the heat exchangers are designed to meet these specifications if possible.

The analysis begins by setting a fixed turbine inlet temperature for all turbines. Then an exit temperature is chosen for all turbines. This generates an estimated compressor pressure ratio (CPR). State points for both the Brayton and Rankine cycles are then calculated. The code iterates on the compressor ratio to deliver an exit air pressure slightly above atmospheric, given the temperature constraints. In a subiteration, it calculates the ratio of mass flows in the Rankine and Brayton cycles.

Once the cycle state points have been identified and the pressure ratio converged, the output from the cycle is normalized to the desired power level. This sets the total mass flows of air and water. Once the mass flow rates have been calculated, it is possible to size all of the components. The compressor and turbines are sized based on a correlation with state-of-the-art components and simple scaling rules. Each of the heat exchangers is designed based on the configuration chosen from the Kays and London text [18,24].

Finally, the volumes of all the components are summed to get an estimate of system size.

To optimize the efficiency of the combined cycle, there are two main parameters that must be considered. These are the gas turbines' exit temperatures and the steam cycle peak pressure. The peak gas turbine inlet temperature is set as an input parameter. It can be varied, but it is obvious that the higher the gas turbine inlet temperature, the better the efficiency will be. The same can be said for most of the components. If their efficiency is higher, the efficiency of the cycle will be higher. The same is true of atmospheric conditions. The colder the input air and the circulating water, the better the efficiency will be. Therefore, most of the input parameters are chosen based on nominal values. The real variables are the turbine exit temperatures and the steam pressure in the bottoming cycle. It is not obvious though what these parameters should be to achieve peak efficiency. They must be varied to identify the peak efficiency achievable.

Note that the turbine exit temperatures that affect the heat exchanger return temperatures to the reactor cannot totally be set without reference to the reactor inlet temperatures. However, due to the high heat transfer characteristics of a molten salt–cooled reactor, or a liquid metal–cooled reactor, there is enough design flexibility to decouple this effect.

The analysis for the recuperated system is much simpler than the combined cycle analysis because the steam cycle does not have to be modeled. The CPR iteration is much simpler, but still sets the pressure ratio so as to meet an exit pressure slightly above atmospheric. The new calculation in the recuperated system is for the air-to-air recuperator. As the recuperator will be the largest component in the system, it does not make sense to prescribe anything other than a counterflow heat exchanger. In this case, the pressure drops for the hot and cold fluids cannot be set independently for simple heat transfer correlations. The pressure drop on the hot air side was chosen as the flow path setting parameter. In this case, also, most of the parameters in the code models are set to nominal values. The only parameter that must be varied to optimize the efficiency of the system is the gas turbines' exit temperatures. The choice of the exit temperature is made to achieve the peak efficiency.

5.6 Validation of Methodology

Before proceeding to estimate the performance of advanced systems, it is useful to validate the methodology by estimating the performance of a currently deployed system. By 2000, GE had over 893 combined cycle and cogeneration systems installed worldwide producing more than 67,397 MW(electric) (Ref. 19). One of their systems was chosen to model. The system chosen was an S107FA 60-Hz 250-MW gas turbine with a three-pressure reheat steam cycle. It is a single-shaft system and its performance parameters are given in Table 5.1. GE classes it as a third-generation system and the first installation was in 1995. It has two steam-reheat cycles and three steam pressures. It burns natural gas.

The parameters in Table 5.1 are adequate to calculate the overall efficiency if turbine and compressor polytropic efficiencies are estimated, and an assumption is made for the pressure drop across the combustor.

Polytropic efficiencies were estimated based on the correlations given in Wilson and Korakianitis (Ref. 5, pp. 79–80 and Ref. 9, Table 6.2, p. 363), giving 94% for the compressor and 93% for the turbine. The standard 5% pressure drop across the combustion chamber was assumed.

Table 5.1 GAE's S107FA performance parameter [19].

Compressor pressure ratio	14:1
Firing temperature	1588 K
Turbine entrance temperature	1422 K
Steam boiler pressure	1400–1800 psi
Steam turbine exhaust pressure	1.2 inch Hg
Power	250 MW (electric)
Site conditions	59 °F, 14.7 psi, 60% Rh
Overall efficiency	56.5%

With these data and assuming an ideal cycle (neglecting the added fuel mass in the turbine, which would add slightly to the turbine output), the methodology used here gives an overall efficiency of 55.8–56.3% depending on the pressure in the bottoming steam cycle. Thus, our methodology underestimates the overall efficiency by 0.2–0.7%. This is certainly adequate for a preliminary calculation.

It should be pointed out that the simple model for the efficiency given by Eq. (5.4) based on a Brayton cycle efficiency of 41.2% and a Rankine cycle efficiency of 42.3% (both calculated) gave an efficiency of

$$\eta_T = \eta_B + \eta_R - \eta_B \eta_R$$
$$= 0.412 + 0.423 - 0.412 \times 0.423$$
$$= 0.661$$

It overestimates the combined cycle efficiency by 9.6%. The detailed model used for this analysis does significantly better.

5.7 Modeling the Nuclear Combined Cycle

Having demonstrated that our analytic approach predicts a reasonable efficiency for current gas turbine combined cycle systems, it can be used to predict the performance of NACC systems and NARC systems.

A nominal set of conditions was chosen as a best estimate for environmental conditions and component performance. A peak turbine inlet temperature of 933 K was chosen as the baseline condition. It is anticipated that this will be achievable by both the molten salt reactors and the liquid lead or lead-bismuth reactors. The high-temperature gas reactors will do better, but helium-to-air heat exchangers have not been considered here. For the combined cycle system, the number of turbines, the turbine exit temperature, and the steam pressure in the bottoming cycle were varied to achieve the maximum thermodynamic efficiency [25]. After the optimum efficiency was determined, the sensitivity of this result to important parameters was estimated. For the recuperated cycle systems, the number of turbines and the turbine exit temperatures were varied to achieve the maximum efficiency. The nominal input parameters for the combined cycle systems (NACC) are given in Table 5.2.

Table 5.2 Combined cycle system baseline.

Number of turbines	Varied (1–5)
Turbine inlet temperature	933 K
Turbine exit temperature	Varied (730–860 K)
Turbine polytropic efficiency	0.90
Compressor pressure ratio	Calculated
Compressor polytropic efficiency	0.90
Main heater pressure ratios	0.99
Atmospheric pressure	1 atm
Atmospheric temperature	288 K (15 °C)
Circulating water input temperature	288 K (15 °C)
Ratio of exhaust pressure to atmospheric	0.98
Air pressure ratios across each superheater	0.99
Air pressure ratio across the evaporator	0.99
Air pressure ratio across the economizer	0.99
Pinch point temperature difference	10 K
Temperature difference of steam exit from superheaters	15 K
Peak Rankine cycle pressure	Varied (1–12 MPa)
Intermediate Rankine cycle pressure	Varied, ¼ of peak
Low Rankine cycle pressure	Varied, 1/16 of peak
Condenser pressure	9 kPa
Steam turbine thermal efficiency	0.90
Power level	25 MW (electric)

5.7.1 Nominal Results for Combined Cycle Model

The turbine exit temperatures and the peak pressure in the steam bottoming cycle were varied for systems using one to five turbines. The best efficiency achievable in each case is plotted in Fig. 5.6. The efficiency is a monotonic function of the number of turbines, with the five-turbine system only slightly better than the four-turbine system. So, the four-turbine case was chosen as the baseline representative combined cycle system. The peak efficiency for the four-turbine case is 45.9% and for the five-turbine case is 46.5%. The optimum turbine exit temperature for the four-turbine case is 810 K and the best steam pressure in the bottoming cycle was 3 MPa. The detailed results are presented in Fig. 5.7. The system underperforms with a steam pressure of only 1 MPa in the bottoming cycle, but from 2 to 4 MPa, the results are very similar. The optimum exhaust temperature shifts slightly from 800 to 810 K.

The CPR versus turbine exit temperature is plotted in Fig. 5.8. The CPR that achieves the peak efficiency is identified by the dotted line. As the exhaust temperature and the steam pressure in the bottoming cycle

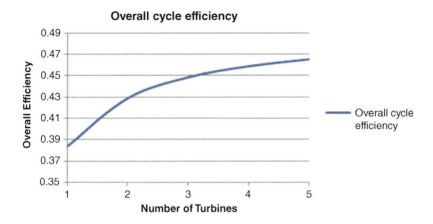

Fig. 5.6 Peak overall thermal efficiency versus number of turbines.

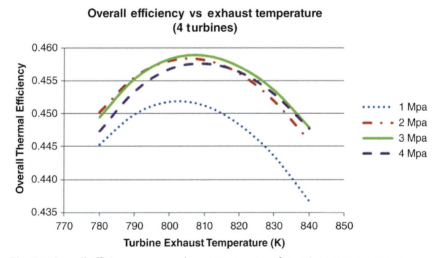

Fig. 5.7 Overall efficiency versus exhaust temperature for various steam pressures.

vary in a discrete manner in fairly large steps, the calculated points jump when switching from two to three turbines.

Two major sensitivities were considered: compressor polytropic efficiency and pressure drops (or ratios) in the main heaters. Fig. 5.9 presents the dependence of the overall thermal efficiency on the gas compressor polytropic efficiency. The overall thermal efficiency increases about 0.4% for every 1% increase in the compressor polytropic efficiency. The pressure drops in the main heaters are analogous to the pressure drop in a

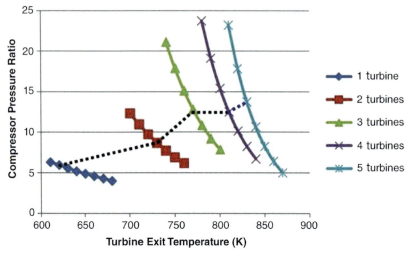

Fig. 5.8 Optimum pressure ratio versus number of turbines.

combustion chamber. Nominally the pressure drop in each heater was set to 1%, but the effect of these pressure drops on the overall efficiency was estimated.

Fig. 5.10 describes this effect. The overall efficiency drops approximately 0.5% for every 1% increase in the pressure drop in the main heater for the four-turbine system.

The main purpose of designing the heat exchangers and estimating sizes for pump, compressor, and turbine components was to get an overall estimate of the size for the complete power conversion system. This estimate

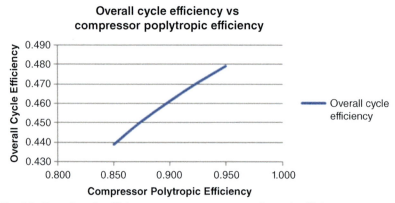

Fig. 5.9 Overall cycle efficiency versus compressor polytropic efficiency.

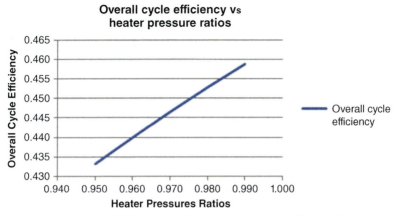

Fig. 5.10 Overall cycle efficiency versus heater pressure ratio (four turbines).

is provided in Fig. 5.11. The system volume consists of the volumes of the air–Brayton system and the steam–Rankine system. The air–Brayton volume includes the volumes of the compressors, turbines, and primary heat exchangers.

The volumes of the compressors and turbines are based on correlations derived from aero-derivative engines as a function of mass flow rate and pressure ratios. The volumes of the primary heat exchangers are calculated based on data from Ref. 18. The steam-Rankine volume includes the steam turbines, pump, superheaters, evaporator, economizer, and condenser. The steam turbines are based on a correlation developed as a function of mass

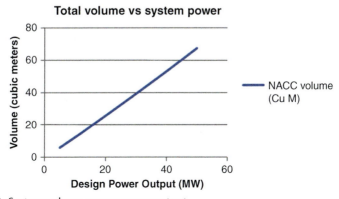

Fig. 5.11 System volume versus power output.

flow rate, inlet steam density, and pressure ratio. The volumes of the superheaters, evaporator, economizer, and condenser are calculated based on the data from Ref. 18.

The pump volume is neglected, and the total volume for both systems in multiplied by a factor of 1.2 to take into account the interconnections.

Note that for the 25-MW(electric) system the air turbines produce 57% of the power and the steam turbines produce 43% of the power. The air cycle has an efficiency of 26.3% and the steam cycle has a thermal efficiency of 35.5%. The steam cycle extracts 30.3 MW of thermal energy and must reject 19.5 MW of waste heat.

A traditional steam plant operating with a thermal efficiency of 38% and producing 25 MW(electric) would have to reject 40.8 MW of thermal energy. The NACC plant cuts the circulating water requirement by slightly over 50%. The steam bottoming cycle fills 77% of the estimated volume and the air topping cycle fills 23% of the volume. These percentages appear to remain relatively constant with increasing or decreasing power levels.

5.7.2 Extension of Results for Peak Turbine Temperatures

The turbine inlet temperature of 933 K chosen for this study is aggressive but within the range projected for the molten salt reactor and the lead or lead-bismuth–cooled reactor. Should it be possible to achieve even higher temperatures in the future, we estimated the efficiencies that might be achieved. Fig. 5.12 gives the anticipated efficiencies that can be achieved by a nuclear combined cycle system for up to about 1100 K turbine inlet temperature.

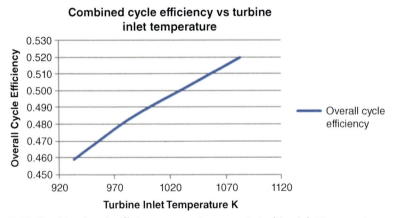

Fig. 5.12 Combined cycle efficiency versus increases in turbine inlet temperature.

5.8 Modeling the Nuclear Recuperated Cycle

The NARC was modeled in a manner similar to that for the combined cycle system to give a direct comparison. The following section gives a description of this modeling and the results obtained.

5.8.1 Nominal Results for Recuperated Cycle Models

For this case, one to five turbines were considered also. However, the only parameter of interest for optimizing the overall cycle efficiency is the turbine exhaust temperature. The results for all five turbine models are presented in Fig. 5.13. For the recuperated Brayton cycles, the efficiency peaks with three turbines under the nominal conditions, but the peak efficiencies for two and four turbines are very close. The actual numbers are given in Table 5.3. So, all sensitivity studies were done on the three-turbine model, and it is interesting to note that the peak efficiency achieved for the recuperated system of 44.6% is only,1.3% less than the peak efficiency obtained for the combined cycle system.

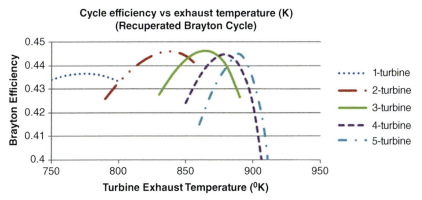

Fig. 5.13 Recuperated Brayton cycles versus turbine exhaust temperatures.

Table 5.3 Peak cycle efficiencies for recuperated Brayton system.

Turbines	Peak efficiency (%)	Pressure ratio	Turbine exit temperature (K)	Mass flow rate (kg/s)
1	43.7	2.27	780	298.3
2	44.5	2.63	840	241.5
3	44.6	3.09	860	207.9
4	44.5	2.99	880	208.3
5	44.2	3.05	890	211.4

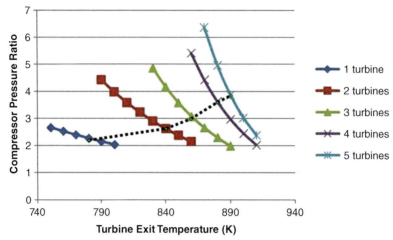

Fig. 5.14 Compressor pressure ratios for peak efficiency versus number of turbines.

The CPRs as a function of the turbine exit temperatures are plotted in Fig. 5.14. The dotted line connects the optimum efficiency pressure ratio. As expected for a recuperated system, the CPRs are significantly less than those for an unrecuperated system. The baseline system is described in Table 5.3.

The sensitivity of the cycle efficiency to recuperator pressure drops and recuperator effectiveness were estimated and are presented in Fig. 5.15. The cycle efficiency drops 0.6% for every 1% increase in the recuperator pressure drop. This is one of the more sensitive parameters because the pressure drops occur on both the cold leg and the hot leg, and the CPR has to compensate for both of them.

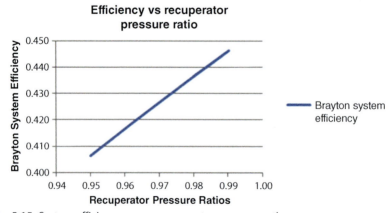

Fig. 5.15 System efficiency versus recuperator pressure ratio.

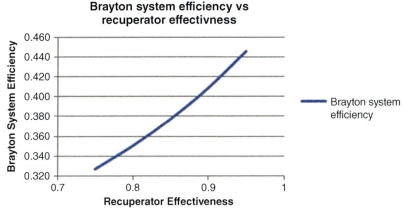

Fig. 5.16 Efficiency versus recuperator effectiveness (three turbines).

The recuperator effectiveness of 0.95 for the baseline cases produces a very large recuperator, so the effectiveness of the recuperator was varied to determine its effect on the overall cycle efficiency and the volume of the resulting system. The effectiveness of the recuperator's impact on the overall efficiency is plotted in Fig. 5.16.

The overall system efficiency also drops approximately 0.6% for every 1% decrease in the recuperator's effectiveness.

The recuperator volume estimates for the baseline 25-MW system are plotted in Fig. 5.17. It appears that asking for an effectiveness of 0.95 causes a very significant size increase compared to an effectiveness of 0.9. Giving

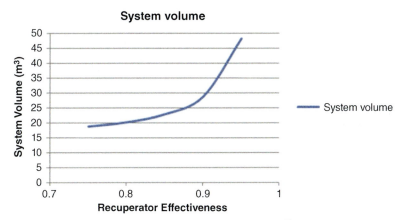

Fig. 5.17 Recuperated system volume versus recuperator effectiveness.

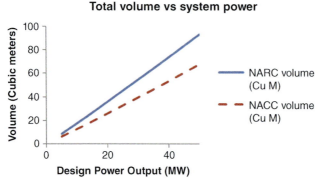

Fig. 5.18 System volume versus system power for the recuperator system.

up 5% in effectiveness causes a 4% loss in system efficiency but reduces the size of the system by 40%.

Fig. 5.18 gives the recuperated system size as a function of total system power for the range from 5 to 50 MW. The recuperated system requires a significantly larger (about 40%) volume than the combined cycle system and produces a slightly lower efficiency, but it has fewer components and is free of the circulating water requirements.

5.8.2 Nominal Results for Recuperated Cycle

The turbine inlet temperature of 933 K chosen for this study was aggressive but within the range projected for the molten salt reactor and the lead or lead-bismuth–cooled reactor. As it may be possible to achieve even higher temperatures in the future, we estimated the efficiencies that might be achieved. Fig. 5.19 gives the anticipated efficiencies that can be achieved

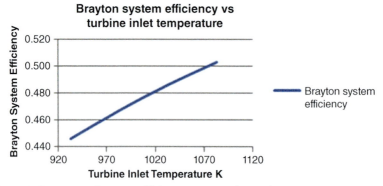

Fig. 5.19 Recuperated system efficiency versus turbine inlet temperature.

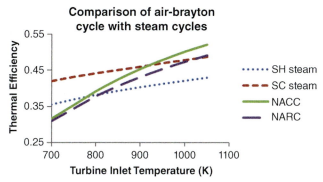

Fig. 5.20 Comparison of NACC and steam cycles.

by a nuclear open air recuperated cycle for up to about 1100 K turbine inlet temperature. Remember that these temperatures are still about 500 K below state-of-the-art gas turbine inlet temperatures.

A summary comparison is made in Fig. 5.20 of the NACC and NARC systems with two steam cycles. The NACC system reaches the efficiency of an extrapolated superheated steam cycle [7] at about 800 K and the efficiency of an extrapolated supercritical (SC) water cycle [20] at about 950 K. Of course, both of these temperatures are well above the normal maximums for the steam cycles but certainly accessible by an NACC or NARC system. For an SC water reactor, a pressure of 25 MPa and an outlet temperature of 500 °C are projected. The estimated efficiency is about 43.8% [20]. Both the NACC and NARC systems exceed this at 933 K. Current superheat technology tops out at about the same temperature and 10–12.5 MPa. The NACC system has a peak pressure in the air system of 1.25 MPa and 3–4 MPa in the steam system. The NARC system has a peak pressure of 0.32 MPa. The lower pressures are less stressing on the systems at their peak temperatures. If higher temperatures are sought to increase system efficiencies, then it is clear that the air-Brayton systems will outperform steam systems.

5.9 Economic Impact

The economic impact of developing an NACC or NARC system over current LWR steam systems, which are trapped in the vapor dome, will come from the following three effects:

(1) The overall thermal efficiency will be higher, saving fuel costs
(2) The components will be smaller and possibly less expensive based on the large ancillary market
(3) More sites will be accessible due to the lowered circulating water requirements.

Table 5.4 Fuel and nonfuel costs of nuclear power (miles/kWh).

	France	Japan	Canada	United States[a]	United States[b]
Nonfuel costs	13.5	21.5	29.8	36.7	25.5
Fuel Costs	7.3	10.1	3.1	7.1	7.1

[a]Average duration of construction.
[b]Shortest duration of Construction.

The Department of Energy sponsors cost guidelines based on detailed economic studies. [21] Unfortunately they tend to go out of date fairly rapidly in a free economy, so no attempt will be made here to do a quantitative calculation. However, some qualitative comments can be made.

The effect of improved efficiency for a nuclear system will never be as significant as it is for a chemical system, but it is still significant. Consider Table 5.4, which was extracted from Knief [22]. An increase in thermal efficiency from the current 30% to 45% will reduce fuel costs by 33%. This will reduce the cost of electricity in France by 12%, in Japan by 11%, in Canada by 3%, and in the United States by 5.4–7.3%.

These savings are significant and will be more so as fuel expenses become a larger fraction of the overall costs of operating a nuclear plant with longer plant lifetimes.

If the nuclear power conversion market can take advantage of the gas turbine combined cycle market and the aero-derivative gas generator market, the costs of components should be significantly reduced. Quantifying this reduction is very difficult, but at this time there appear to be far more incentives to improve the efficiency of gas generators and combined cycle components than there do to improve the efficiency of large nuclear steam generators and turbines.

Finally, the lowered requirement for circulating water to get rid of waste heat will have an impact in several ways. First, the components, such as cooling towers, can be significantly reduced in size or eliminated. Second, the decreased need for cooling water will open up many more sites for installation of nuclear systems. NARC systems can be installed anywhere. This will reduce transmission costs as power sources can be located closer to their loads. Of course, with an NARC system, a large recuperator will be required to replace the cooling towers. But recuperators lend themselves to assembly line construction, and shipment to site, much better than do cooling towers.

Clearly the economic incentives are more qualitative than quantitative, but they are real.

5.10 Conclusions

The analysis performed here does an excellent job of modeling combined cycle plants and allows prediction of performance based on a number of plant design features. It predicts plant efficiencies above 45% for reactor temperatures predicted to be in the 7008 °C (973 K) range. It also predicts a reduction in circulating water requirements by over 50% compared to current generation reactors.

The circulating water requirements can be eliminated completely if a recuperator is used instead of the bottoming steam cycle. The efficiency penalty for going to a recuperated system appears to be only about 1–2%. The size of the system increases by about 50%, however. But many more sites become accessible to power stations, possibly reducing transmission costs.

No attempt to analyze a gas-cooled reactor was undertaken in this study. Given that efficiencies will continue to increase with even higher reactor outlet temperatures, it could prove fruitful to consider this possibility in the future. The primary heaters will be large, but they will operate at high pressure so they will not grow like the recuperator does. This brings up another possibility for improving a recuperated system's performance. It might be possible to exhaust at higher pressure and reduce the size of a 0.95 effective recuperator. Little has been done in this study to optimize the recuperated design. Future work may well be able to reduce the size of the recuperator and keep the high efficiency.

Though this study focused on a 25-MW(electric) power conversion system, it only considered one loop. Typically, an LWR system has three or four loops, and so the results of this study could be easily extended to a 100-MW(electric) plant. This should cover the range for most of the Generation IV SMRs that are being considered. There is also talk of replicating SMRs to build up to a significant power station. The 25-MW(electric) system components appear to be a reasonable point to start hardware development.

System efficiencies around 45% appear very achievable with conservative performance for the various heat exchangers required for each of these systems. Two- and three-turbine systems appear to achieve performances comparable to a four-turbine system. Reheat cycles in the bottoming steam cycle do not appear to be advantageous at the pressures (10 MPa) currently being used in combined cycles. Lower pressures of the order of 2–3 MPa appear to reverse this result and let efficiencies improve with steam reheat cycles.

It would appear that the analysis accomplished here only scratched the surface of the possibilities for this type of system and thermal efficiencies easily above 45% should be readily achievable.

References

[1] J. H. Horlock Combined Power Plants, Including Combined Cycle Gas Turbine (CCGT) Plants. Krieger Publishing Company, Malabar, FL (2002).
[2] J. H. Horlock, Cogeneration-Combined Heat and Power (CHP), Thermodynamics and Economics. Krieger Publishing Company, Malabar, FL (1997).
[3] R. Kehlhofer, et al., Combined Cycle Gas & Steam Turbine Power Plants, Third ed., PennWell, Tulsa, OK, 2009.
[4] B. Meherwan, Gas Turbine Engineering Handbook, Second ed., Gulf Professional Publishing, Houston, TX, 2002. ISBN 0 88415732 6.
[5] D.G. Wilson, T. Korakianitis, The Design of High-Efficiency Turbomachinery and Gas Turbines, Second ed., Prentice-Hall Inc., Upper Saddle River, NJ, 1998.
[6] P.P. Walsh, P. Fletcher, Gas Turbine Performance, Blackwell Science Ltd and ASME, Fairfield, NJ, 1998.
[7] M.M. El-Wakil, Powerplant Technology, McGraw-Hill Book Company, NY, 1984.
[8] L.O. Tomlinson, S. McCullough, Single Shaft Combined Cycle Power Generation System, GER-3767C, General Electric Power Systems (1996).
[9] J.D. Mattingly, Elements of Gas Turbine Propulsion, McGraw-Hill Inc., New York, 1996.
[10] G.C. Oates, Aerothermodynamics of Gas Turbine and Rocket Propulsion, American Institute of Aeronautics and Astronautics, Washington, DC, 1988.
[11] A.P. Fraas, Heat Exchanger Design, Second ed., John Wiley & Sons, New York, 1989, pp. 350–354.
[12] P.F. Peterson, Multiple-reheat Brayton cycles for nuclear power conversion with molten coolants, Nucl. Technol. 144 (2003) 279, http://dx.doi.org/10.13182/ NT144-279.
[13] C. Forsberg, D. Curtis, Meeting the needs of a nuclear-renewables electrical grid with a fluoride-salt–cooled high-temperature reactor coupled to a nuclear air-Brayton combined-cycle power system, Nucl. Technol. 185 (2014) 281, doi:10.13182/NT13-58 http://dx.doi.org/.
[14] V. Dostal, et al., Medium-power lead-alloy fast reactor balance-of-plant options, Nucl. Technol. 147 (2004) 388, doi:10.13182/NT147-388 http://dx.doi.org/.
[15] R.E. MacPherson, J.C. AMOS, H.W. SAVAGE, Development testing of liquid metal and molten salt heat exchangers, Nucl. Sci. Eng. 8 (1960) 14, doi:10.13182/NSE8-1-14 http://dx.doi.org/.
[16] A.P. Fraas, Design precepts for high-temperature heat exchangers, Nucl. Sci. Eng. 8 (1960) 21, doi:10.13182/NSE8-1-21 http://dx.doi.org/.
[17] M.M. Yarosh, Evaluation of the performance of liquid metal and molten salt heat exchangers, Nucl. Sci. Eng. 8 (1960) 32, doi:10.13182/NSE8-1-32 http://dx.doi.org/.
[18] W.M. Kays, A.L. London, Compact Heat Exchangers, McGraw-Hill Book Company, NY, 1964.
[19] D.L. Chase, Combined Cycle Development Evolution and Future GER-4206, General Electric Power Systems. (2001) http://sites.science.oregonstate.edu/~hetheriw/ energy/topics/doc/elec/natgas/cc/combined%20cycle%20development%20evolution%20and%20future%20GER4206.pdf.
[20] Y. Oka, S. Koshizuka, Design Concept of Once-Through Cycle Super-Critical Light Water Cooled Reactors, Proceedings of International Symposium of Supercritical Water-Cooled Reactors (SCR 2000), 2000 November 6–8.

[21] J. G. Delene and C. R. Hudson II, Cost Estimates Guidelines for Advanced Nuclear Power Technologies, ORNL/TM-10071/R3, Oak Ridge National Laboratory (May 1993).
[22] R.A. Knief, Nuclear Engineering, Theory and Technology of Commercial Nuclear Power, XXXX, La Grange Park, Illinois, 2008, pp. 225 Table 8-3American Nuclear Society.
[23] B. Zohuri, Dimensional Analysis and Self-Similarity Methods for Engineers and Scientists, First ed, Springer Publishing Company, NY, 2017.
[24] B. Zohuri, Compact Heat Exchangers: Selection, Application, Design and Evaluation, First ed, Springer Publishing Company, NY, 2017.
[25] B. Zohuri, P.J. McDaniel, Thermodynamics in Nuclear Power Plant Systems, Second ed, Springer Publishing Company, NY, 2019.

CHAPTER 6

Heat pipe driven heat exchangers to avoid salt freezing and control tritium

6.1 Introduction

Heat transfer from the primary salt coolant in a fluoride-salt-cooled high-temperature reactor (FHR), molten salt reactor (MSR), or salt-cooled fusion reactor presents two challenges not seen in water-, sodium-, and helium-cooled reactors:
1. Liquid salts have melting points above 400 °C and must not freeze to avoid blocking of primary coolant flow.
2. The salt coolant generates tritium that must not be allowed to escape to the environment. Lithium fluoride is used in coolant salts to lower the melting point of the coolant salt; however, neutron irradiation of lithium generates tritium. Isotopically separated ^{7}Li is used in salt fission systems to minimize neutron adsorption and tritium production.

Isotopically separated ^{6}Li is used in fusion systems to maximize tritium production because tritium is the fuel for a fusion reactor. Tritium production rates in salt-cooled fusion machines are about three orders of magnitude larger than in fission machines.

This chapter examines the use of heat exchangers that incorporate multiple heat pipes designed to:
1. transfer heat from primary coolant to power cycle, secondary loop, or environment;
2. provide the safety function of a secondary loop by isolating the reactor salt coolant from the high-pressure power cycle;
3. stop heat transfer if the reactor coolant approaches its freezing point to prevent flow blockage; and
4. block tritium escape to the environment with recovery of the tritium.

No such heat pipe with all of these capabilities exists but each of these four capabilities has been demonstrated in heat pipes. If heat pipes can be developed with these four capabilities, there is the potential to greatly simplify salt reactor design and reduce costs by replacing multiple independent systems with a single system that meets the four requirements.

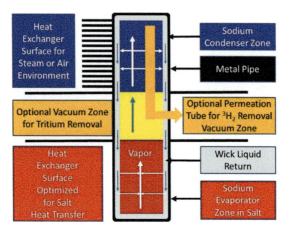

Fig. 6.1 Heat transfer regions within a heat pipe.

We define the requirements for such heat pipes, review the status of the technology to meet the four functional requirements, and define the path forward. Fig. 6.1 shows a schematic of a sodium heat pipe system with these capabilities.

As discussed later, sodium is the preferred heat pipe fluid for salt systems based on multiple criteria [23,37,58].

6.2 Heat transfer—the traditional application for heat pipes

The bottom half of the heat pipe is in the hot salt. The heat vaporizes the sodium that then flows to the colder condenser section of the heat pipe where it condenses and flows back to the hot salt zone. Heat pipes operate at near isothermal conditions where the rate of heat transfer is determined by the rate of sodium vapor flow from evaporator to condenser zone. As the temperature goes up, the density of sodium vapor increases, and more heat is transferred by the evaporation–condensation cycle. The heat pipe is lined with a wick where capillary forces wet the entire heat pipe surface and move liquid sodium from the cold condenser section back to the hot salt section, where it is evaporated. The heat pipe operates independent of orientation of the heat pipe. If there were a pool of sodium, the boiling point of the sodium would be greater at the bottom than farther up the heated zone due to the hydrostatic pressure and would require higher temperature drops across the heat pipe for the same heat transfer rate. The use of wicks minimizes the heat pipe sodium inventory and thus the consequences of a failed heat pipe [23, 58].

6.3 Prevention of coolant salt freezing

Heat pipes can be designed to start up at a preset temperature. Sodium vapor pressure and thus sodium heat transfer rapidly increase with temperature. The sodium atmospheric boiling point is at 883 °C. In a sodium heat pipe, the temperature needs to be above 500 °C for significant heat transfer—above the freezing points of salt coolants. Second, one can put a small amount of inert gas into the heat pipe. At low sodium pressure, this gas blocks sodium vapor flow to the condenser section. As the temperature increases, the sodium pressure increases, the inert gas is compressed into smaller gas volume at the top of the heat pipe and most of the condenser section is available for sodium vapor condensation [23,37,58].

6.4 Tritium capture

Heat pipes can be efficient methods for tritium capture. Most of the surface area in any nuclear system is associated with the heat exchangers. Unless one has methods for rapid removal of tritium ($^3H_2$) from the salt and diffusion barriers on the surface of the heat exchangers, the tritium will rapidly diffuse through the heat exchangers. With a heat pipe, the sodium condenser section on the inside is coated with tungsten, aluminum oxide, or other material to limit tritium escape via diffusion through the heat pipe. Liquid salts are fluxing agents that dissolve oxides and most high-performance tritium diffusion barriers. In contrast, sodium is chemically compatible with many high-performance tritium diffusion barriers including aluminum-oxide diffusion barriers. The different chemical environment enables a much wider choice of tritium diffusion barriers. The sodium vapor sweeps the tritium gas to the condenser section to be removed by a nickel permeation tube where the tritium diffuses from the sodium vapor to the nickel surface, diffuses through the nickel, and is removed from a vacuum zone on the other side. This tritium removal system can be small relative to other tritium removal systems because of four system characteristics: (1) the large heat exchanger area enables tritium to diffuse through the salt to the heat pipe surface, (2) noncondensable tritium in the heat pipe is concentrated in the condenser section by sodium vapor movement, (3) tritium diffusion through sodium vapor and sodium liquid to the permeation membrane is rapid relative to the slow diffusion of tritium through liquid salts, and (4) tritium diffusion through the nickel is rapid. Mechanisms number (2) and (3) create the potential of a tritium removal system that is an order of magnitude smaller than alternative methods for tritium removal from salt systems.

Heat pipes are used today for a wide variety of applications from electronics to building heating to space nuclear reactors [58]. Recent work with sodium heat pipes includes a 2018 ground demonstration of a nuclear space reactor with sodium heat pipes [1] to move heat from the reactor to the power cycle. Today sodium heat pipes are being considered for microreactors [2,59] with power ratings up to 15 MW(electric). The interest in microreactors implies development of heat exchangers at the multimegawatt scale using sodium heat pipes—the same base technology that would be used for salt-cooled fission and fusion reactors with the added considerations of (1) heat transfer shutdown as liquid salts approach their melting points to avoid freezing salt and (2) control of tritium.

Sections 6.5 and 6.6 describe alternative fission and fusion salt systems and the requirements for these heat exchangers. Section 6.7 provides a general description of heat pipe operation and how to use heat pipes to prevent salt freezing. Section 6.8 discusses control of tritium and other chemical aspects of the design.

6.5 Reactor systems and heat pipes design requirements

There are multiple salt fission and fusion systems that will impose somewhat different requirements on heat pipe systems [23, 37, 58].

6.5.1 Fluoride-salt-cooled high-temperature reactor

The FHR (Refs. 3,4, and 5) uses graphite matrix-coated particle fuel developed for high-temperature gas-cooled reactors (HTGRs) and clean liquid fluoride salt coolants originally developed for MSRs. The concept is less than 20 years old and no FHR has been built. Fluoride salt coolants are used because of their chemical compatibility with the graphite fuel and the low neutron adsorption cross-section of fluorine. Relative to high-temperature gas-cooled reactors, the use of a liquid salt coolant enables (1) increasing the power density by up to a factor of 10 because liquids are better coolants than gases, (2) near atmospheric pressure operation, and (3) a more robust safety case based on the fuel and a coolant that dissolves fission products that escape the fuel.

The pebble-bed FHR (Fig. 6.2B) is being developed by Kairos Power that uses flibe ($^7Li_2BeF^4$) salt because of its excellent neutronic and thermal–hydraulic properties [6–8] relative to other salts (Table 6.1). Isotopically separated ^7Li is required because of the very high neutron adsorption cross section of ^6Li. However, even with the use of isotopically separated 7Li, the

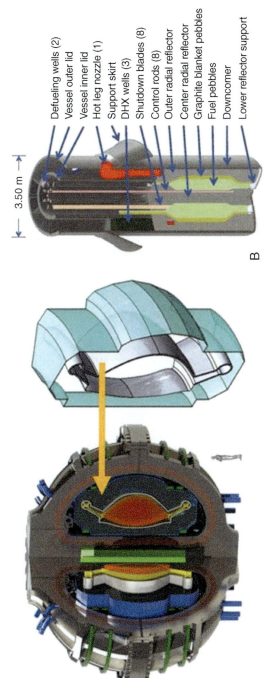

Fig. 6.2 (A) Affordable, robust, compact (ARC) fusion system and flibe fusion blanket for ARC and (B) FHR fission system.

Table 6.1 Candidate fluoride-salt reactor coolants*.

Coolant	T_{melt} (°C)	T_{boil} (°C)	ρ (kg/m³)	$\rho C p$ (kJ/m³ °C)
Li$_2$BeF$_4$ (flibe)	459	1430	1940	4670
59.5 NaF–40.5 ZrF$_4$	500	1290	3140	3670
26 ^{7}LiF–37 NaF–37 ZrF$_4$	436	–	2790	3500
51 ^{7}LiF–49 ZrF$_4$	509	–	3090	3750
Water (7.5 MPa)	0	290	732	4040
Water (15.5 MPa)	0	345	709	4049

* References [6,7,8]. Salt compositions in mole percent. Salt properties at 700 °C and 1 atm. The boiling points of the zirconium fluoride salts are not well known. For comparison, water data are shown at 290 °C (7.5 MPa, boiling point) and pressurized water reactor conditions with water properties at 309 °C.

tritium generation rate is about two orders of magnitude larger than in a pressurized water reactor. Methods to capture the tritium and prevent its release to the environment are required.

These salts will typically deliver heat to the power cycle at between 600 °C and 700 °C. The minimum salt temperature is significantly above the melting point of flibe (459 °C) for two reasons: (1) flibe near its melting point is viscous and (2) a reasonable temperature margin is required to avoid the risk of freezing salt in the heat exchangers. The maximum salt temperature is controlled by the availability of economic materials of construction. With existing materials that limit is near 700 °C.

For the FHR, the allowable fuel and coolant temperature limits could allow peak coolant temperatures to near 900 °C assuming suitable materials of construction for the heat exchangers [60].

6.5.2 Salt-cooled fusion systems

Clean (no-fuel) flibe salt blankets are being developed for high-magnetic-field fusion machines. In a fusion machine radioactive tritium (^{3}H$_2$) combines with deuterium (^{2}H$_2$) to generate helium and 14-MeV neutrons. Fig. 6.2A shows the fusion system and flibe blanket. The flibe salt has four functions:
1. It slows down neutrons and converts that energy into heat.
2. The neutrons are used to breed tritium.
3. The salt acts as the primary radiation shielding to protect the magnets.
4. Flibe salt cools the first wall that separates the fusion plasma from the salt. Flibe salt is chosen to maximize tritium production.

The beryllium in the salt acts as a neutron multiplier via a (n, 2n) reaction. Neutron adsorption by lithium generates tritium. Current designs propose 90% ^{6}Li in the salt to maximize tritium production resulting in tritium production rates about three orders of magnitude greater than in an FHR per unit of thermal power.

While the concept of a flibe fusion blanket is old, recent engineering developments have created large incentives to develop flibe salt blankets relative to another fusion blanket concepts. In the last 5 years, advances in superconductors have enabled doubling the magnetic field in fusion reactors [9,10]. In fusion systems, the plasma power density increases as one over the fourth power of the magnetic field. Higher magnetic fields can reduce the machine volume by an order of magnitude for the same total power output, which dramatically improves the economics but creates major challenges. In a fusion blanket, the 14-MeV neutrons slow down, delivering most of the heat from the fusion reactor to the blanket. With these higher power densities, it is difficult to cool traditional solid fusion blankets. As a consequence, flibe liquid-salt immersion blankets are proposed to adsorb the heat from the fast neutrons and breed tritium fuel. The first of these high-magnetic-field fusion systems is being developed by Commonwealth Fusion Systems in cooperation with the Massachusetts Institute of Technology. One consequence of these developments is that these advanced fission and fusion systems are tightly coupled [11] in terms of salt coolant technology, tritium control, and power cycle development.

6.5.3 Molten salt reactors

MSRs dissolve the fuel in the salt. Fluoride salts are used in thermal spectrum MSRs, whereas fluoride or chloride salts are used in fast spectrum MSRs. Fission products are generated in the liquid salt. Most of the fission products form fluoride or chloride salts. Gaseous volatile fission products (xenon, krypton, etc.) must be removed, and much of the tritium will be removed from the salt at the same time.

The 8-MW (thermal) Molten Salt Reactor Experiment at Oak Ridge National Laboratory demonstrated the concept in the late 1960s. The Shanghai Institute of Applied Physics of the Chinese Academy of Sciences is building a 2-MW (thermal) MSR. Both of these reactors are thermal spectrum MSRs with a graphite moderator that uses flibe salt with isotopically separated ^{7}Li. Some thermal spectrum MSR designs use other fluoride salts.

Most but not all MSRs propose that in an emergency the liquid fuel drain to a critically safe, passively cooled dump tank, typically using a freeze valve of frozen salt where if the reactor heats up, the salt melts, and the salt drains to the dump tanks [12]. In China [13–15], work is underway to develop sodium-cooled heat pipes to transfer decay heat from this dump tank to the air. This includes ongoing laboratory experiments. For this application, the heat pipe requirements are relaxed relative to the FHR and fusion applications—tritium generation stops once the reactor becomes noncritical and the decay heat is decreasing with time. There is also ongoing work to develop potassium-cooled heat pipes for decay heat removal in FHRs (Ref. [16]). Neither of these systems are designed for tritium recovery or temperature control.

Europeans [17–19] are developing a fast spectrum MSR using fluoride salts with a composition of 77.5% LiF and 22.5% heavy metals that includes thorium and the fissile fluorides. The salt mixture requires lithium fluoride to reduce the melting point of the liquid salt. Isotopically separated ^{7}Li is used to minimize tritium production, but as with the FHR there will be significant tritium production.

Last, there is the recent development of molten chloride fast reactors. Only fission product tritium is generated in these systems, which do not include lithium in the chloride salts. The chloride salts are typically mixtures of sodium, potassium, and sometimes magnesium salts. The optimum salt compositions are not fully defined for these systems. The molten chloride fast reactor [20] is being developed by TerraPower as a breed-and-burn once through MSR. Because there is no lithium or beryllium, tritium production will be less than in most other MSR systems, with much of the tritium stripped from the salt with the xenon and krypton gases. In this specific design, the liquid salt remains in the main system when shutdown; dump tanks are not used.

The other molten chloride fast-reactor concept, being developed by Moltex [21-22], has the traditional layout of a solid-fuel fast reactor except the fuel pins contain liquid chloride salts. A secondary clean salt transfers heat from the fuel assemblies to the heat exchanger. In this context, it resembles the FHR with fixed fuel, a clean liquid salt coolant, and a requirement to avoid salt freezing.

6.6 Salt reactor heat exchanger requirements

Heat exchangers will be used to transfer heat (1) from the primary coolant to the power cycle and (2) remove decay heat after shutdown for the FHR

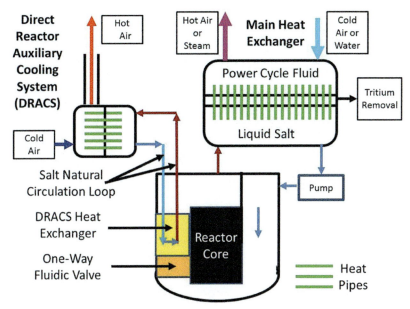

Fig. 6.3 FHR with primary and decay heat removal heat exchangers incorporating heat pipes.

and MSR systems. A simplified schematic of an FHR with these two types of heat exchangers is shown in Fig. 6.3 using heat pipe heat exchangers.

The primary difference between salt reactors and other reactors (water, helium, or sodium cooling) in the context of heat exchangers is that the salt melting point is above 400 °C; thus, a system is required to prevent freezing of the salt within the heat exchanger. The choices are (1) a power cycle where the cold incoming fluid is above the salt melting point, (2) a control system that isolates the heat exchanger or dumps the salt if salt temperatures approach the freezing point of the salt, or (3) a heat exchanger that shuts down before the salt can freeze—a heat pipe heat exchanger.

The primary coolant heat exchangers will send heat to a Rankine (steam) or nuclear air-Brayton power cycle (see Chapter 5 of this book). Within that heat exchanger multiple heat pipes (green in Fig. 6.3) move heat from the salt to the steam cycle through a tube sheet. There is the option to include a tritium removal system as a component of the heat pipe system. Heat pipes are nearly isothermal heat transfer devices. For transferring heat from the salt to the power cycle, countercurrent heat exchanges are preferred to maximize the temperature of the steam or compressed air to the power cycle. Fig. 6.4 shows schematically how one may use multiple

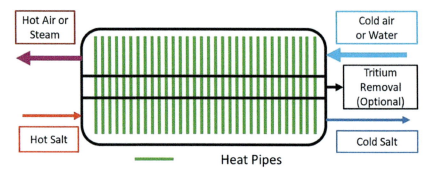

Fig. 6.4 Countercurrent heat exchanger with heat pipe.

heat pipes to create a countercurrent heat exchanger. Countercurrent heat pipe heat exchangers are used in heat recovery systems [23–28].

A more advanced power cycle option is the nuclear air-Brayton combined cycle that has peaking capability [3–5, 29–32]. MSRs were originally developed as part of the U.S. Aircraft Nuclear Propulsion Program to power a jet bomber of unlimited range; that is, salt reactors were originally developed to efficiently couple to a Brayton power cycle with a sodium intermediate loop between the reactor and jet engine.

The Brayton power cycle requirements drove the reactor design (see Chapter 5 and Appendix A of this book). Because of changes in the electricity market with the addition of wind and solar and advances in gas turbine technology, there are now large incentives to couple salt reactors to gas turbine cycles because of their capabilities to produce peak power using thermodynamic topping cycles as a replacement for the traditional gas turbine in a low-carbon electricity grid. One candidate gas turbine cycle is shown in Fig. 6.5. Air is compressed, heated using nuclear heat, goes through a turbine, is reheated using nuclear heat, goes through a second turbine, and then goes to a heat recovery steam generator. It is a nuclear variant of a natural gas combined cycle.

Added peak power can be produced by adding high-temperature heat or a combustible fuel after the nuclear reheat step to increase the gas temperature going into the second turbine up to temperatures as high as 1500 °C. The combustible fuel could be natural gas, biofuels, or ultimately, hydrogen. If there is low-price electricity, the electricity can be used to heat firebrick (firebrick resistance-heated energy storage) that then can provide very high-temperature heat for the peaking cycle. The peak power capability is a thermodynamic topping cycle with incremental heat-to-electricity efficiency that can approach 75%, and the cycles can be designed

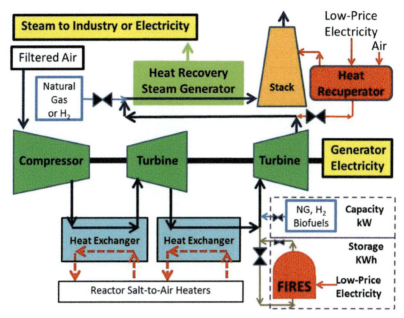

Fig. 6.5 Nuclear air–Brayton combined cycle with peaking capability.
Note: For further understanding of direct reactor auxiliary cooling system (DRACS), please see Appendix B of this book.

for different base load–to–peak power ratios to match market requirements, including large peak power capabilities to provide large assured generating capacities.

The temperature exiting the front-end air compressor of a modern gas turbine is near 400 °C. That requires the reactor to provide heat to the power cycle between 500 °C and 700 °C and matches salt reactors. The high compressed air temperatures reduce the risk of freezing the salt. However, because it is an open-air gas turbine, it implies that any tritium that goes through the heat exchanger will be released to the environment in seconds.

The use of heat pipes avoids concerns about freezing of the salt when the reactor shuts down and can potentially block tritium releases to the environment. Because of the air pressure in the gas turbine, any heat pipe failure will have the air entering the heat pipe, the oxygen reacting with sodium to form sodium oxide, and a resulting shutdown heat pipe filled primarily with nitrogen. There are other implications. The air-side heat transfer coefficient is less than in steam systems; thus, these heat exchangers are considerably larger than salt-to-steam-system heat exchangers.

The FHR has a passive decay heat cooling system where the goals are to minimize heat losses during normal operations, remove heat if the salt coolant temperature increases such as after reactor shutdown, and avoid freezing the salt. Several alternative systems [33,34] have been proposed to meet these goals. A direct reactor auxiliary cooling system (DRACS) (see Appendix B) with heat pipes can be used as shown in Fig. 6.3. DRACS is a natural circulation heat transfer system using flibe or another salt that transfers heat from the primary system. In this specific case, a sodium heat pipe would include an inert gas that would minimize heat to the environment when the salt temperature was below its nominal peak operating temperature.

The DRACS heat exchanger (DHX) in the primary system is located between the lower higher-pressure cold salt plenum below the reactor core and upper hot salt plenum above the reactor core. When the pumps are running, the flow through the DHX is restricted by a one-way fluidic valve that allows some bypass flow through the DHX from the cold-salt zone at the bottom of the reactor to the hot zone above the reactor core. If the pump stops, hot salt exiting the reactor flows through the DHX to the bottom of the reactor core by natural circulation. This heats the salt in the separate DRACS loop that flows by natural circulation to the salt–air heat exchanger where the heat is dumped to air. The air flows by this heat exchanger by natural circulation.

In the salt-to-air heat exchanger, heat pipes move heat from the natural circulation salt to the air stream. The heat pipes turn on and off when salt temperatures are above or below 700 °C, respectively. A recent assessment [35] concluded that in an accident the peak temperature must not exceed 970 °C to ensure against structural failure of the 316 stainless steel. This defines the performance requirements for DRACS. The high-temperature fuel and coolant capabilities imply that temperature limits are defined by components outside the reactor core, such as heat exchangers, piping, and the reactor vessel.

Last, heat pipes can be used to cool hot-box systems. Rather than insulate individual pipes, valves, pump heads, sample lines, salt inventory tanks, primary reactor vessel, and other components to prevent salt freezing, there is the option to insulate a box that contains multiple such components.

There is less risk of small tubes freezing, inspection is simplified, and total insulation costs may be reduced. For such hot boxes, heat pipes can be used to dump heat from the hot box to the atmosphere if the salt temperatures and thus hot-box room temperatures exceed some temperature limit. Unlike the above cases, in this case the heat pipe evaporator section [23] would be in an inert gas environment rather than in a liquid salt environment. Heat transfer to the heat pipe would be some combination of convective gas flow and radiative heat transfer. The heat pipe could be part of the wall structure.

6.7 Heat pipe design and startup temperature

In this section, we will expand on heat infrastructure technology and design from holistic point of view with few pages in Appendix C of this book as well, however more granular information can be found in Zohuri book (Ref. [23]).

6.7.1 Choice of fluid

Different salt reactors have somewhat different salt inlet and outlet temperatures, but within the temperature range of 500 °C–750 °C. Advanced machines may operate at higher temperatures. Decay heat cooling systems may also operate at higher temperatures. The candidate heat pipe fluids are potassium, sodium, and lithium as shown in Table 6.2.

There has been large-scale development of sodium, potassium, and lithium heat pipes [36–40], primarily for space nuclear reactors. Many of these studies and experiments have included start up and shut down under hot and cold temperatures. Sodium, as discussed below, is the preferred heat transfer fluid for most salt systems.

The operating range for each heat pipe fluid is determined by the liquid metal vapor pressure versus pressure. At very low pressures, there is insufficient flow of vapor from the evaporator to condenser section to move significant heat. As discussed below, in most heat pipe systems the maximum heat transfer is a function of the density of the vapor and the sonic velocity of the vapor.

Fig. 6.6 shows the vapor pressure of sodium [41] and potassium as a function of temperature. A potassium heat pipe becomes fully operational substantially below the freezing point of liquid salts [40] with significant

Table 6.2 Heat pipe coolant options [23].

Fluid	Boiling point (°C) at 1 atm	Typical operating range (°C)
Lithium (Li)	1330	825–1125
Sodium (Na)	883	600–825
Potassium (K)	759	325–525

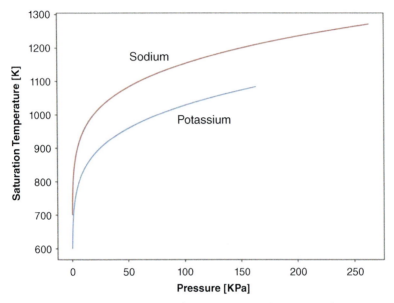

Fig. 6.6 Vapor pressure (in kilopascals) of sodium (top) and potassium (bottom) versus temperature (in kelvins) [23].

potassium pressures at higher temperatures. Sodium heat pipes begin to transfer heat at higher temperatures [1] as the sodium vapor pressure increases. There is substantial experimental data on heat pipe performance with temperature, primarily with space reactors that go through a startup transient from frozen liquid metal to full operating temperatures. The minimum heat transfer in a heat pipe as the temperature goes down is determined by heat conduction through the heat pipe walls and wick structure between the evaporator zone and condenser zone that contains liquid sodium.

The startup temperature of the heat pipe can be more precisely controlled by the addition of an inert gas to the heat pipe (Fig. 6.7). For sodium heat pipes the inert gas is neon, which has about the same molecular weight as sodium. See Appendix C of this book as well.

The flow of sodium vapor from hot to cold pushes the inert gas to the top of the condenser section. At lower temperatures and, thus lower sodium pressures, the inert gas fills the condenser section with very little heat flow. As the temperature increases, the sodium vapor pressure increases, and the flow of sodium vapor pushes the inert gas to the top of the heat pipe. When this happens more of the condenser section is exposed to sodium vapor and

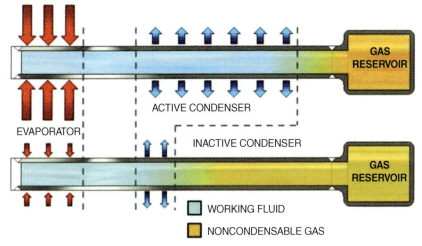

Fig. 6.7 Schematic of variable conductance heat pipe [23].

is condensed. The amount of inert gas for any given geometry determines at what temperature the boundary between the inert gas and the sodium vapor is pushed into the condenser section. This creates the option of a decay heat removal system with low heat losses during normal reactor operations, high heat removal if the temperature goes up, and shut down as temperatures go down. Experiments [42] have been done with sodium heat pipes with neon gas for temperature control.

This heat pipe startup and shutdown capability is being considered for concentrated solar power (CSP) towers. Concentrated light from the sun provides the heat to the evaporator section of the heat pipe. The heat is transferred to flowing liquid salt. Liquid salts are used for heat storage. Sodium heat pipes have two potential advantages. First, sodium is very good at transferring heat with very high heat fluxes to the salt. The peak temperature of currently used nitrate salts is limited by the potential for hot spots in the solar receiver. The heat pipe eliminates the potential for excess nitrate temperatures and allows somewhat higher average peak temperatures. Second, the addition of an inert gas can be used to shut down the heat pipe to avoid salt freezing. When a cloud comes over the CSP system, the heat pipe will operate in reverse to cool down and then freeze the salt. The current generation of CSP systems uses nitrate salt with a relatively low melting point, but the next generation of CSP systems is expected to use a sodium–potassium–magnesium–chloride salt that can operate at very high temperatures and is very cheap. However, that salt has a melting point of approximately 420 °C. A heat pipe with the appropriate design can become a one-way heat transfer system to prevent salt freezing—the same challenge as with salt-cooled reactors.

The operating temperature of a sodium heat pipe can be raised—adding lithium to the sodium heat pipes will lower the vapor pressure of the sodium at any given temperature. We are not aware of any experimental work that has investigated this option for sodium heat pipes. Dual-component heat pipes [43] have been investigated with other fluids but have more complicated behavior. In some ways, their operation is similar to a distillation column on full reflux. The lower boiling point component preferentially vaporizes resulting in the gas-phase composition changing with temperatures. In addition, molten lithium complexes with tritium—a way to capture tritium. However, in such a heat pipe there may be no simple way to recover the tritium. This is discussed in Section 6.9.

The performance of heat pipes is measured by the Merit number as defined in Eq. (6.1) and plotted in Fig. 6.8 as a function of temperature for different fluids [23]:

$$N_l = \frac{\rho_l \sigma \lambda}{\mu_l} \qquad (6.1)$$

where ρ_l is the liquid density, σ is the surface tension, λ is the latent heat, and μ_l is the liquid viscosity.

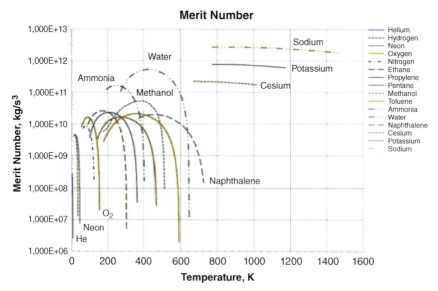

Fig. 6.8 Merit number for commonly used heat pipe working fluids [23].

High liquid density and high latent heat reduce the fluid flow required to transport a given power, while high surface tension increases the pumping capability. A low liquid viscosity reduces the liquid pressure drop for a given power. Sodium has the highest Merit number for any heat pipe system; that is, the best performance.

The combination of operating temperature range and performance indicates that a sodium-based heat pipe system is the preferred option for salt systems.

6.8 Heat transfer analysis

We examine herein the limits of heat pipes and then the implications for sodium heat pipes.

6.8.1 Heat pipe operation limits

Heat pipes have multiple operational limits but only a few of these limits apply to a sodium heat pipe in a liquid salt system. Fig. 6.9 shows the operational limits for a specific sodium heat pipe versus temperature and the constraints that are likely to be relevant for fission or fusion salt system—sonic (gas phase) and capillary (liquid phase) limits.

We discuss all the limits and why specific limits are important or unimportant for sodium heat pipes in salt systems. This specific heat pipe has an inside diameter of 1.5 cm and a length of 4 m. Under normal conditions (salt reactor conditions), such a heat pipe would transfer kilowatts to tens of kilowatts of heat. For near-term salt reactor systems, normal peak operating temperatures would be near 700 °C.

A computer run for designing heat pipe limit for MSR operating under normal peak temperature is illustrated in Fig. 6.10.

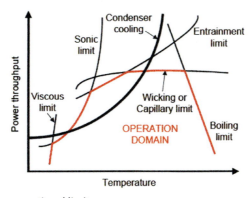

Fig. 6.9 Heat pipe operational limits.

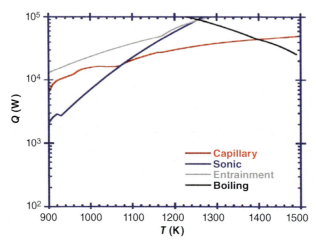

Fig. 6.10 Typical pipe axial performance limits versus evaporator exit temperature for 1.5 cm inside diameter by 4-m-long sodium heat pipe [44].

Fig. 6.10 is a typical run for a fixed heat pipe design and may be different than a variable heat pipe [44].

6.8.1.1 Viscous limit

The viscous limit dominates at low temperature, near the melting point of the working fluid. The high liquid pressure losses in the wick, due to high viscosity and low permeability, limit liquid flow from the condenser to the evaporator section. Salt melting points in salt reactors are far above the melting point of sodium so this is not a constraint for sodium heat pipes in salt systems [23].

The viscous limit derivation is shown in book by Zohuri [23] and is written in the following equation form as:

$$Q_{vapor_{max}} = \frac{\pi r_v^4 h_{fg} \rho_{v_e} P_{v_e}}{12 \mu_{v_e} l_{eff}} \tag{6.2}$$

where P_{v_e} is the vapor pressure in evaporator (Pa), ρ_{v_e} is the vapor density in evaporator (kg/m³), r_v is the cross-sectional radius of the vapor core (m), α_{v_e} is the vapor viscosity in the evaporator (N s/m²), l_{eff} is the effective length of heat pipe at the lower end of the working-fluid temperature range (m), and L is the latent heat of vaporization (J/kg).

6.8.1.2 Entrainment limit

The entrainment limit is a result of the liquid and vapor moving in opposite directions. It influences wick design. The vapor exerts a shearing force on the liquid at the liquid–vapor interface. If this shear force exceeds the surface tension of the liquid, liquid droplets are entrained into the vapor flow and are carried toward the condenser section. The entrainment limit is typically encountered during a heat pipe startup when the vapor flow at the evaporator section exit is choked (velocity is near sonic).

Entrainment limit equation is presented here and the derivation of it can be found in book by Zohuri [23].

$$Q_{e_{max}} = A_v \lambda \left(\frac{\sigma \rho_v}{2 r_{h,s}} \right)^{1/2} \quad (6.3)$$

where ρ_v is the vapor density (kg/m^3), $r_{h,s}$ is the surface pore hydraulic radius (m), A_v is the vapor core cross-section area (m^2), σ is the surface tension coefficient (N/m), and λ is the latent heat of vaporization (J/kg).

The entrainment limit can be raised by employing a small pore size wick and/or increasing the cross-sectional flow area for the vapor in the heat pipe to lower its velocity at the exit of the evaporator section. This would not be expected in a properly designed sodium heat pipe [23].

6.8.1.3 Boiling limit

Boiling at the inside surface of the heat pipe wall in the evaporator section is likely when the local liquid superheat exceeds that for incipient nucleate boiling. The ensuing nucleation and growth of vapor bubbles blocks the flow of returning liquid to the evaporator section. In alkali-metal heat pipes, the boiling limit is typically encountered at high wall temperatures, beyond those selected for nominal operation.

Equation of Boiling limit is presented in the book by Zohuri [23] and presented here as well.

$$Q_{b_{max}} = \frac{2\pi L_e k_e T_v}{\lambda \rho_v \ln(r_i / r_v)} \left(\frac{2\sigma}{r_n} - P_c \right) \quad (6.4)$$

where $Q_{b_{max}}$ is the boiling heat transport limit (Watts), L_e is the evaporator section length (m), k_e is the effective thermal conductivity of the liquid or saturated wick matrix (W/m K), T_v is the vapor temperature (K), λ is the latent heat of vaporization (J/kg), ρ_v is the vapor density (kg/m^2), r_i is the inside radius of pipe (m), r_v is the vapor core radius (m), σ is the surface tension coefficient (N/m), r_n is the boiling nucleation radius (m), and P_c is the capillary pressure (N/m^2).

The effective thermal conductivity, K_e, used in Eq. (D-9) is highly dependent upon the wick geometry [61]. Chi gives the equation for finding the effective thermal conductivity of an axially grooved heat pipe.

$$k_e = \frac{wk_l(0.185w_f k_w + \delta k_l) + (w_f k_l k_w \delta)}{(w + w_f)(0.185w_f k_w + \delta k_l)} \quad (6.5)$$

where w_f is the Groove fin thickness (m), w is the Groove width (m), δ is the Groove depth (m), k_l is the liquid thermal conductivity (W/m K), and k_w is the wall thermal conductivity (W/m K).

There are other form of this equation under different condition also derived in Ref. [23].

Boiling limits can occur in heat pipe reactors, where the evaporator section of the heat pipe is in the reactor core. For sodium heat pipes in salt systems, where liquid salt brings the heat to the heat pipe, this is not a design constraint. Liquid salts have a low Prandtl number; that is, low thermal conductivity that limits heat transfer rates. The heat transfer limits on the salt side of the evaporator are far more limiting than the limits inside the heat pipe [23].

6.8.1.4 Sonic limit

The sonic limit is dominant at lower temperatures for many designs of sodium heat pipes (see Fig. 6.9). The vapor pressure of the working fluid is a good indicator of reaching this limit [45]. The vapor pressure and physical state of the heat pipe liquid at ambient temperature, as well as the thermal resistance between the condenser and the adjacent heat sink, have significant influence of the startup behavior of a heat pipe. The sonic limitation is defined in Eq. (6.6):

$$Q_{s,\max} = A_v \rho_v h_{lv} \left[\frac{\gamma_v T_v r_v}{2(\gamma_v + 1)} \right]^{1/2} \quad (6.6)$$

where A_v is the cross-sectional area of vapor space, ρ_v is the vapor density phase (kg/m^3) h_{lv} is the heat transfer coefficient liquid, vapor (latent heat of vaporization), γ_v is the specific heat ratio for vapor side, r_v is the radius of vapor space section, and h_{lv} is the heat transfer coefficient liquid, vapor.

6.8.1.5 Wicking or capillary limit

The wicking limit or capillary limit can be a constraint at higher temperatures for sodium heat pipes. This condition occurs when an applied heat flux causes the liquid in the wick structure to evaporate faster than it can be supplied by the capillary pumping power of the wick. Once this event takes place the meniscus at the liquid–vapor interface continues to withdraw and move back into the wick until all of the liquid has been depleted.

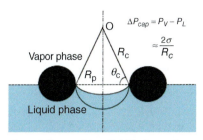

Fig. 6.11 Capillary pumping of the working fluid in heat pipe.

This action will cause the wick to become dry and the heat pipe container temperature may continue to rise at the evaporator until a "burnout" condition is reached [45]. The difference in the capillary pressure across the liquid–vapor interfaces governs the operation of the heat pipes. This is one of the most important parameters that affect the performance and operation of a heat pipe [23].

The capillary limit is encountered when the capillary pressure is not sufficient to pump the liquid back to the evaporator and thus causes the dry out of the wick of the evaporator end. The physical structure of the wick is one of the most important reasons for this limit, and the type of working fluid affects it. It is less of a concern for vertical heat pipes where gravity supports the return of the liquid to the evaporator section. [23]

Thus, the capillary (or wicking) limit is encountered when the net capillary pressure head is less than the combined pressure losses of the liquid flow in the wick and of the countercurrent vapor flow in the heat pipe. The capillary pressure head for circulating the heat pipe's working fluid increases with increasing the liquid surface tension and decreasing the radius of curvature of the liquid–vapor meniscus in the surface pores in the wick, as illustrated in terms of R_c in Fig. 6.11 and presented by Eq. (6.7), as *capillary limitation*.

$$Q_{c,\max} = \frac{2\left(\dfrac{\sigma \rho_l h_{lv}}{\mu_l}\right)\left(\dfrac{K}{r_c}\right)(2\pi r_v T_w)}{0.5 L_e + L_a + 0.5 L_c} \qquad (6.7)$$

where σ is the surface tension, condensing coefficient (N/m), ρl is the liquid density phase (kg/m^3), h_{lv} is the heat transfer coefficient liquid, vapor (latent heat of vaporization), K is the permeability (m^2), r_v is the radius of vapor space section, t_w is the temperate of heat pipe wall, μ_l is the liquid viscosity (N s/m^2), r_c is the effective capillary radius (m), L_e is the length of the evaporator section (m), L_a is the length of the adiabatic section (m), and L_c is the length of the condenser section (m).

6.8.2 Sodium heat pipe experience

There is significant experience with sodium heat pipes in the temperature range applicable to salt reactor systems. Thermacore, Inc., has carried out several sodium heat pipe life tests to establish long-term operating reliability. A 316 L stainless steel heat pipe with a sintered porous nickel wick structure and an integral brazed cartridge heater has successfully operated at 650 °C–700 °C for over 115,000 h without signs of failure. A second 316 L stainless steel heat pipe with a specially designed Inconel 601 rupture disk and a sintered nickel powder wick has demonstrated over 83,000 h at 600 °C–650 °C with similar success. A representative one-tenth-segment Stirling Space Power Converter heat pipe with an Inconel 718 envelope and a stainless steel screen wick has operated for over 41,000 h at nearly 700 °C. A hybrid (i.e., gas fired and solar) heat pipe with a Haynes 230 envelope and a sintered porous nickel wick structure was operated for about 20,000 h at nearly 700 °C without signs of degradation.

These life test results collectively have demonstrated the potential for high-temperature sodium heat pipes at the temperatures expected in salt fission and fusion systems to serve as reliable energy conversion system components for power applications that require a long operating lifetime with high reliability [39]. There have been many sodium heat pipes built and tested for space applications including long-duration runs (>100,000 h) and different operating temperatures. In a space reactor, heat pipes move heat from the reactor around the radiation shield to the power conversion system.

Multiple heat pipes are used rather than a typical sodium heat transfer loop so any single failure (tube leak) does not cause mission failure. This experience includes testing heat pipes in the Experimental Breeder Reactor II for space applications [38, 45], where heat would be transferred at temperatures ranging from 900 °C to 1100 °C [37, 58].

Heat pipes can be designed to be large or small. The maximum heat pipe size will likely be determined by the allowable leakage of sodium into the primary system if there is a leak in the heat pipe. That, in turn, determines the total number of heat pipes required for each application.

Sodium is a strong chemical reducing agent that will reduce beryllium fluoride and many other fluorides that may be salt components [46] of the coolant salt to metals. Most of these metals have some solubility in the coolant salt. Small additions of sodium to the primary coolant will not have major impacts; but large additions would cause major changes in salt chemistry with different impacts depending upon the type of salt-cooled system.

These parameters will place upper limits on individual heat pipe sodium inventories to limit the consequences of leaks. In this context, the use of a wick is important because it minimizes the sodium inventory in any individual heat pipe.

6.9 Tritium control

For many fission salt-cooled reactors and all salt-cooled fusion reactors, neutron interaction with the coolant generates tritium. Fission reactors use isotopically separated ^{7}Li to minimize tritium production whereas fusion machines use isotopically separated ^{6}Li to maximize tritium production. Tritium is the fuel for fusion reactors. The tritium production rates in FHRs per megawatt hour are about hundred times those of a pressurized water reactor. Tritium production rates in a fusion machine will be a thousand times larger per unit of power output than an FHR. Further, fusion machines will have the added requirement for efficient recovery of tritium to recycle as fuel while having the same allowable release limits to the environment.

Most of the tritium is generated by neutron adsorption by lithium in lithium fluoride; thus, the chemical form of the tritium is ^{3}HF. All proposed salt reactors have chemical redox control systems [46] to convert corrosive ^{3}HF into ^{3}H that then becomes ^{3}H$_2$ or ^{3}HH if hydrogen gas sparging is used for tritium removal. Tritium becomes a dissolved gas (^{3}H$_2$) in the liquid salt. To capture tritium and prevent its release to the environment, the fission or fusion reactor will contain barriers to tritium release and methods to capture the tritium. Recent reviews [47] have examined methods to capture tritium and limit its release from salt-cooled fission and fusion reactors.

Three strategies are available. Real systems are likely to use more than one system, particularly fusion reactors with their very high tritium production rates:

1. *Tritium removal from salt*: Tritium can be removed from liquid salts by beds of carbon and other materials, permeation filters where the tritium diffuses through nickel or other tubing to a tritium removal system, and gas sparging using inert gases or gases that contain some hydrogen for redox control.
2. *Tritium removal within heat exchangers*: The options herein are heat pipes and double-wall heat exchangers. Double-wall heat exchangers require

a vacuum or flowing fluid between the double tubes for tritium removal with some loss in heat exchanger efficiency.
3. *Tritium removal in a secondary loop*: If a nitrate salt intermediate heat transfer loop is used, tritium that diffuses through the heat exchangers from the hot salt will be converted to steam and can be removed from the off-gas of the nitrate salt system.

Heat pipes can incorporate (1) barriers to prevent the release of tritium from the primary system and/or (2) a tritium capture system. Materials with low permeability to tritium, such as tungsten, can be used to coat the salt side (evaporator section) of the heat exchanger to minimize tritium diffusion into the heat pipe. Alternatively, tungsten, aluminum oxide, or other coatings can be used to coat the inside sodium condenser side of the heat pipe to not allow tritium to escape from the heat pipe to the environment. Tritium diffusion barriers on the inside of the sodium heat pipe have the potential to be orders of magnitude better than tritium diffusion barriers in salt systems. Liquid salts act as fluxing agents that dissolve almost all proposed tritium diffusion barriers except options such as tungsten. However, sodium is chemically compatible with high-performance tritium diffusion barriers, such as aluminum oxide, that can be applied to metal heat pipes.

Tritium can be removed from the heat pipe by a permeation filter where tritium diffuses into a vacuum or inert sweep gas, as shown in Fig. 6.1. This can be part of the wall of the heat pipe or a separate permeation filter can be installed in the heat pipe. In each case there is a vacuum to recover the tritium. The sodium vapor will sweep the tritium that diffuses through the evaporator section toward the condenser section of the heat pipe and thus most of the surface area of the permeation filter should be in the condenser section. Tritium will diffuse through the metallic permeation filter into the vacuum zone. Nickel is the likely material of construction for the permeation tube due to its compatibility with sodium and high permeability to hydrogen.

There have been many studies on removing tritium from liquid salts with permeation membranes [48]. Permeation rates are a function of the hydrogen diffusivity and hydrogen solubility in the material. Fig. 6.12 shows hydrogen diffusivity for different materials that may be found in an FHR or fusion reactor including salts, metals of construction, and liquid sodium.

Fig. 6.13 shows the corresponding solubility coefficients for different materials that could be found in an FHR or fusion machine with heat pipes. There is a caveat on hydrogen solubility. For diatomic gas molecules, the solubility in metals, including liquid sodium, follows Sievert's law where

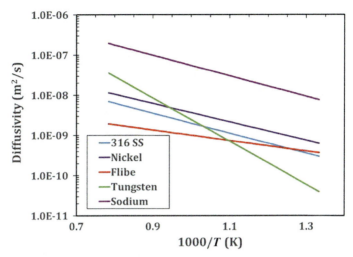

Fig. 6.12 Diffusivity of flibe compared to different metals [42–44] and liquid sodium [45].

the solubility is proportional to the square root of pressure. For liquid salts, the solubility of gas follows Henry's law, where the solubility is proportional to the pressure.

In terms of permeation rates, diffusivity can be considered a measure of the difficulty of a molecule moving through a material whereas the solubility is a measure of the number of pathways for a molecule to move through a material. Liquid salts have low diffusivity relative to most materials and low

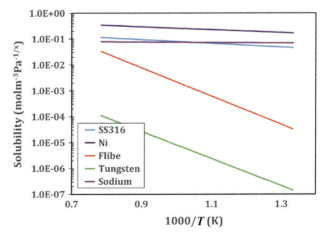

Fig. 6.13 Hydrogen solubility coefficients versus temperature for different materials including liquid sodium.

hydrogen solubility. This implies that any system to remove tritium from the salt must have a high surface area to reduce the diffusion time from the liquid to the removal system, be it a gas sparging system, an absorber bed, or a permeation filter. If a permeation filter is used for removal of dissolved tritium in the salt, it will most likely be like a cross-flow heat exchanger to create turbulent flow of salt past the tubes to increase the mass transfer of tritium from salt to the permeation filter surface. Once the tritium reaches the surface, it can diffuse through the metal for collection of the tritium.

This mass transfer limitation implies large surface areas for the permeation filter. There is, however, one structure in all salt systems with a large surface area and thus the potential to easily remove tritium from the salt—the primary heat exchanger. If it does not have a highly effective permeation barrier, the tritium will go through the heat exchanger.

A permeation filter to remove tritium in a heat pipe will have a much smaller surface area per unit of tritium removed and thus is attractive as a tritium control strategy. The large heat pipe surface area for salt-to-sodium heat transfer implies tritium will easily enter the heat pipe unless the heat pipes have permeation barriers to prevent tritium diffusion from salt to sodium. Inside the heat pipe, the vaporization and condensation of sodium sweeps and concentrates the tritium gas in the condensate section of the heat pipe. The tritium will rapidly diffuse to a permeation membrane tube because of the low density of sodium vapor relative to liquid flibe salt. At 673 °C, the pressure will be a tenth of an atmosphere. This is diffusion through a low density gas about three orders of magnitude less dense than the liquid salt. The combination of using sodium vapor transport to concentrate hydrogen and the low density of sodium vapor relative to salt implies that the surface area of the permeation membrane in the heat pipe is much smaller than a permeation membrane to remove an equivalent amount of tritium from the liquid salt.

The alternative option is having a section of the heat pipe exposed to a vacuum and operating as a permeation membrane (Fig. 6.1). There are two options here. The first option is to have parts of the wall that are used as permeation membranes with no wick material to minimize liquid sodium between the sodium vapor and the wall. The second option is to have sodium in the wick structure flow by the wall to remove tritium from the sodium. Liquid sodium has a high diffusion coefficient (Fig. 6.12) and high solubility (Fig. 6.13) for hydrogen and thus a high permeation rate for hydrogen [52] where the hydrogen dissociates and is atomic hydrogen in the sodium like in solid metals. It is the combination of characteristics that creates the option to use heat pipes as tritium removal systems in a salt fission or fusion reactor. It also follows from Figs. 6.12 to 13 that electroplating tungsten on the inside of the condenser section of the heat pipe will limit tritium escape because of the very low solubility of hydrogen in tungsten. However, as discussed earlier, the sodium environment creates the option to use aluminum oxide or several other high-performance tritium permeation barriers—options that cannot be considered for surfaces exposed to liquid salts that will dissolve such surface barriers.

Hydrogen diffusion into heat pipes and the use of permeation membranes to remove hydrogen has been demonstrated, but not for this application. Heat pipes are being developed for a variety of chemical processes (gasification, solid oxide fuel cell stacks, etc.), where the sodium evaporation (heating) zone is removing heat from hydrogen-rich gas streams many orders of magnitude greater than the tritium in a salt-cooled system. In these systems [53], sufficient hydrogen can diffuse into the heat pipe to create a hydrogen (inert-gas)-flooded condenser zone that shuts down heat transfer in the heat pipe in days. The condenser zone fills with inert hydrogen gas. As a consequence, permeation tubes for hydrogen removal have been developed to remove this hydrogen. Similar systems [54] with permeation filters have been designed and tested for specialized space applications to remove hydrogen where hydrogen diffusion into the heat pipe can shut it down. Some space nuclear reactors use hydrides in the reactor core to moderate neutrons and reduce the reactor size.

These hydrides under some circumstances can lose hydrogen that diffuses into the heat pipes. In all of these systems (1) the quantities of hydrogen are orders of magnitude larger than the quantities of tritium that might enter a heat pipe in a salt reactor system and (2) the permeation of very small amounts of hydrogen through the condenser section of the heat pipe is not a concern. If such systems are used in salt reactor applications, there are tight constraints on allowable tritium escape from the sodium condenser side of the heat pipe system into the environment. We are not aware of any efforts to design and test heat pipes for this combination of constraints and operating conditions.

Tritium will chemically react with sodium to form hydrides [55] at lower temperatures that decompose at higher temperatures. At low concentrations the hydride is soluble in sodium. At very low temperatures, there will be an equilibrium between (1) tritium diffusion into the heat pipe via the evaporator section, (2) tritium forming hydrides, and (3) tritium escaping from the heat pipe via a permeation surface that acts as a tritium removal system or via diffusion from the condenser section of the heat pipe. Lithium forms a much stronger bond with tritium than sodium. If some lithium is mixed with the sodium, one would expect more efficient capture of the tritium. However, any significant lithium will also raise the operating temperatures of the heat pipe. We are not aware of any studies that have undertaken an integrated study of all of the effects and options.

Last, permeation membranes may create other tritium control options. There is the option to use the permeation membrane with a very small surface area to add hydrogen or deuterium at low pressures to the heat pipe to create a hydrogen counter flux through the heat pipe back to the flibe salt to minimize tritium entry into the heat pipe. It is a countercurrent sweep gas. Systems removing the tritium from the salt (gas sparging, absorbers, etc.) would then recover tritium with the added hydrogen and deuterium from the heat pipes. The technical viability of this option depends upon material selection and the flibe surface of the heat pipe. Flibe and some materials such as tungsten have very low solubilities for hydrogen so the addition of a hydrogen/deuterium counter flow can fill the interatomic spaces in the metal or flibe to block tritium diffusion. This is only viable if the required concentration of hydrogen in the heat pipe can be sufficiently low that the hydrogen would not shut down heat pipe operation. To date, there have been insufficient analysis and experimental work to determine if such a system is a technically viable engineering option.

6.10 Status of technology and path forward

Fission and fusion salt systems have two relatively unique challenges relative to other reactor coolants: (1) avoid freezing of the salt and (2) capture the tritium. Heat pipes have the potential to provide a simpler and lower cost solution to these challenges. Some aspects of such heat pipes are well understood but significant work is required in other areas.

Sodium heat pipes for heat transfer are well understood. There is a significant experience base, they have been used in multiple applications, and work is underway to develop larger heat pipe systems for applications including microreactors that couple to different types of power cycles at power levels of tens of megawatts. The larger-scale use of heat pipes is resulting in the development of better models for heat pipe design [56–57], but none of these models are capable of modeling heat pipes with all of the required capabilities as described herein. We are not aware of any work to develop heat pipe heat exchangers for transferring hundreds of megawatts of heat as may be required in a power reactor. The questions are how to design, manufacture, monitor, and repair large heat pipe systems. This is not a constraint if the heat pipe system is used for decay heat removal systems that are much smaller.

Sodium heat pipes that turn off and on at preset temperatures have been built and tested but there is limited operating experience. Most of the work has been associated with understanding the startup behavior of heat pipes, not systems where the temperatures decrease and the system slows down, such as in the long-term behavior of a decay heat removal systems as decay heat decreases with time.

There has been limited work on the performance of heat pipes that capture and remove hydrogen with permeation membranes built into the heat pipe. This work has been focused on systems where hydrogen diffuses into the heat pipe and shuts down the heat pipe; that is, systems where the heat pipe is removing high-temperature heat from a system with high concentrations of hydrogen. In these systems the goal is to keep the heat pipe operating, the hydrogen can be dumped to the environment, and the concentrations of hydrogen are relatively high. We have not identified work beyond very limited studies to develop heat pipes that include tritium recovery where the hydrogen (tritium) concentrations are low and the goals are to (1) capture the tritium and (2) very low tritium losses to the power cycle or environment. At the same time, there have been in the last 5 years rapidly growing efforts to remove and capture tritium from liquid salts, both in experimental work and development of models of tritium transport. Tritium transport to heat pipe surfaces and tritium diffusion through heat exchangers is now reasonably well understood. The science and technology base for a heat pipe to remove and capture tritium is being developed.

References

[1] M. Gibson, et al., "The Kilopower Reactor Using Stirling Technology (KRUSTY) Nuclear Ground Test Results and Lessons Learned" (2019); https://ntrs.nasa.gov/search.jsp? R=201800054352019-04-03T11:44:18+00:00Z. (Accessed 14 July 2019).
[2] A. Levinsky, et al., Westinghouse eVinci TM Reactor for Off-Grid Markets, 119, Trans. American Nuclear Society, Orlando, Florida, 2018.
[3] C.W. Forsberg, P.F. Peterson, Basis for fluoride-salt-cooled high-temperature reactors with nuclear air-Brayton combined cycles and firebrick resistance-heated energy storage, Nucl. Technol. 196 (1) (2016) 13, https://doi.org/10.13182/NT16-28.
[4] C. Andreades, et al., Design summary of the mark-I pebble-bed, fluoride salt–cooled, high-temperature reactor commercial power plant, Nucl. Technol. 195 (3) (2016) 223, https://doi.org/10.13182/NT16-2.
[5] C.W. Forsberg, et al., Integrated FHR Technology Development: Tritium Management, Materials Testing, Salt Chemistry Control, Thermal Hydraulics and Neutronics,

Associated Benchmarking and Commercial Basis, Center for Advanced Nuclear Energy, Massachusetts Institute of Technology, 2018. MIT-ANPTR-180 https://www.osti.gov/search/semantic:1485415. Accessed 14 July 2019.

[6] D.F. Williamas, L.M. Toth, K.T. Clarno, Assessment of Candidate Molten Salt Coolants for the Advanced High-Temperature Reactor (AHTR), Oak Ridge National Laboratory, 2006 ORNL/TM-2006/12.

[7] R.R. Romatoski, L.W. Hu, Fluoride salt coolant properties for nuclear reactor applications: a review, Ann. Nucl. Energy 109 (2017) 635, https://doi.org/10.1016/j.anucene.2017.05.036.

[8] R.R. Romatoski, L.W. Hu, Fluoride-salt-cooled high-temperature test reactor thermal-hydraulic licensing and uncertainty propagation analysis, Nucl. Technol. 205 (2019) 1495, https://doi.org/10.1080/00295450.2019.1610686.

[9] B.N. Sorbom, ARC: a compact, high-field, fusion nuclear science facility and demonstration power plant with demountable magnets, Fusion Eng. Des. 100 (2015) 378 https://www.sciencedirect.com/science/article/pii/S0920379615302337 (Accessed 14 July 2019).

[10] A.Q. Kuang, et al., Conceptual Design Study for Heat Exhaust Management in the ARC Fusion Pilot Plant, Cornell University, 2018 https://arxiv.org/abs/1809.10555. (Accessed 14 July 2019).

[11] C. Forsberg, et al., "Fusion blankets and fluoride-salt-cooled high-temperature reactors with flibe salt coolant: common challenges, tritium control, and opportunities for synergistic development strategies between fission, fusion, and solar salt technologies," Nucl. Technol. Volume 206 (11) (2020): Special issue on Salt-Cooled Reactors; https://doi.org/10.1080/00295450.2019.1691400.

[12] C. Forsberg, Leaky-vessel decay-heat-removal system for fluid-fuel molten salt reactors, Trans. Am. Nucl. Soc. Winter Mtg., Washington, D.C., November 17–21, 2019, 2019.

[13] C. Wang, et al., Transient behavior of the sodium-potassium alloy heat pipe in passive residual heat removal system of molten salt reactor, Prog. Nucl. Energy 68 (2013) 142, https://doi.org/10.1016/j.pnucene.2013.07.001.

[14] C. Wang, et al., Study on the characteristics of the sodium heat pipe in passive residual heat removal system of molten salt reactor, Nucl. Eng. Des. 265 (2013) 691, https://doi.org/10.1016/j.nucengdes.2013.09.023.

[15] M. Liu, et al., Experimental study on the heat transfer characteristics of fluoride salt in the new conceptual passive heat removal system of molten salt reactor, Int. J. Energy Res. 42 (2018) 1635, https://doi.org/10.1002/er.3959.

[16] C. Wang, et al., Conceptual design and analysis of heat pipe cooled silo cooling system for the transportable fluoride-salt-cooled high-temperature reactor, Ann. Nucl. Energy 109 (2017) 458, https://doi.org/10.1016/j.anucene.2017.05.035.

[17] M. Brovchenko, et al., Neutronic benchmark of the molten salt fast reactor in the frame of the EVOL and MARS collaborative projects, EPJ Nucl. Sci. Technol. 5 (2) (2019), https://doi.org/10.1051/epjn/2018052.

[18] E. Merle, "Concept of European Molten Fast Reactor (MSFR)," CNRS-IN2P3-LPSC/Grenoble Institute of Technology/UGA—France (2017); https://www.gen-4.org/gif/upload/docs/application/pdf/2017-05/emerle_gif-final23may2017.pdf (Accessed 14 July 2019).

[19] "SAMOFAR: A Paradigm Shift in Nuclear Reactor Safety with the Molten Salt Fast Reactor" (2019); http://samofar.eu/wp-content/uploads/2019/06/D6.7_SAMOFARDissemination-summary-report_v1.0_20190630.pdf (Accessed 14 July 2019).

[20] B. Feng, et al., "Core and fuel cycle performance of a molten salt fast reactor," in: Presented at ICAPP 2019—International Congress on Advances in Nuclear Power Plants, Juanles-pins, France, May 12–15, 2019.

[21] Advances in Small Modular Reactor Technology Developments, International Atomic Energy Agency, 2018, https://aris.iaea.org/Publications/SMR-Book_2018.pdf Accessed 14 July 2019.

[22] Moltex Energy; https://www.moltexenergy.com/; Supplement to: IAEA Advanced Reactors Information System (ARIS) (2018); https://aris.iaea.org/Publications/SMR-Book_2018.pdf (Accessed 10 January 2020).
[23] B. Zohuri, Heat Pipe Design and Technology: Modern Applications for Practical Thermal Management, first ed., Springer Publishing Company, New York, NY, 2016.
[24] F. Peretz, B. Horbaniuc, Optimal heat pipe heat exchanger design, J. Heat Recov. Syst. 4 (1984) 9, https://doi.org/10.1016/0198-7593(84)90088-2.
[25] X. Zha, S.B. Riffat, G. Gan, Heat recovery with low pressure loss for natural ventilation, Energy Build. 28 (179) (1998), https://doi.org/10.1016/S0378-7788(98)00016-4.
[26] M.S. Ylemez, Optimum length of finned pipe for waste heat recovery, J. Energy Convers. Manage. 49 (2008) 96, https://doi.org/10.1016/j.enconman.2007.05.013.
[27] L.L. Vasiliev, Review heat pipes in modern heat exchangers, J. Therm. Sci. Eng. Appl. 25 (2005) 1, https://doi.org/10.1016/j.applthermaleng.2003.12.004.
[28] E. Azad, Review: heat pipe heat exchangers at IROST, Int. J. Low Carbon Technol. 8 (3) (2013) 173, https://doi.org/10.1093/ijlct/cts012.
[29] B. Zohuri, P. McDaniel, Combined Cycle Driven Efficiency for Next Generation Nuclear Power Plants: An Innovative Design Approach, second ed., Springer, 2018 https://www.springer.com/gp/book/9783319705507 (Accessed 14 July 2019.
[30] B. Zohuri, P. McDaniel, C.R. de Oliveria, Advanced nuclear open air-Brayton cycles for highly efficient power conversion, Nucl. Technol. 192 (1) (2015) 48 https://doi.org/http://dx.doi.10.13182/NT14-42.
[31] N.J. Lamfon, Y. Naijar, M. Akyurt, Modelling and simulation of combined gas turbine engine and heat pipe system for waste heat recovery and utilization, J. Energy Convers. Manage. 39 (81) (1998), https://doi.org/10.1016/S0196-8904(96)00175-6.
[32] A. Greenop, "Coiled Tube Gas Heater Effectiveness Modeling, Simulation and Experiments for Nuclear Power Conversion Cycles," Ph.D. Thesis, University of California Berkeley (2018).
[33] C. Forsberg, Fluoride-Salt-Cooled High-Temperature Reactor (FHR) Temperature Control Options: Removing Decay Heat and Avoiding Salt Freezing, Center for Advanced Nuclear Energy, Massachusetts Institute of Technology, New York, NY, 2019 MIT-ANP-TR-183.
[34] C. Forsberg, et al., Fluoride-salt-cooled high-temperature reactor (FHR) using British advanced gas-cooled reactor (AGR) refueling technology and decay-heat removal systems that prevent salt freezing, Nucl. Technol. 205 (2019) 1127, https://doi.org/10.1080/00295450.2019.1586372.
[35] J. Nakata, et al., Performance evaluation of DRACS system for FHTR and time assessment of operation procedure, Presented at the 21st International Conference on Nuclear Engineering (ICONE21), Chengdu, China, July 29–August 2, 2013.
[36] M.S. El-Genk, J.P. Tournier, Uses of liquid-metal and water heat pipes in space reactor power systems, Front. Heat Pipes (FHP) 2 (2011) 013002, https://doi.org/10.5098/fhp.v2.1.3002.
[37] B. Zohuri, Heat Pipe Applications in Fission Driven Nuclear Power Plants, Springer, 2019, https://www.springer.com/gp/book/9783030058814. (Accessed 14 July 2019).
[38] M.G. Houts, D.I. Poston, W.J. Emrich Jr., Heat pipe power system and heat pipe bimodal system development status, in: M.S. El-Genk (Ed.), Proceedings of Space Technology and Applications of International Forum (STAIF-1998), 3, American Institute of Physics, 1998, pp. 1189–1195 AIP-CP-420.
[39] J.H. Rosenfeld, et al., "An overview of long duration sodium heat pipe tests," Thermacore International, Inc., NASA/TM-2004-212959 (2004); https://ntrs.nasa.gov/archive/NASA/casi.ntrs.nasa.gov/20130013063.pdf (Accessed 10/01/2020).
[40] P.M. Dussinger, W.G. Anderson, E.T. Sunada, Design and testing of titanium/cesium and titanium/potassium heat pipes, advanced cooling technologies, Advanced Colling Technologies (2019) https://www.1-act.com/design-and-testing-oftitaniumcesium-and-titanium potassium-heat-pipes. (Accessed 14 July 2019).

[41] W.H. Rodebush, E.G. Walters, The vapor pressure and vapor density of sodium, J. Am. Chem. Soc. 52 (7) (1930) 2654.

[42] R.S. Reid, J.T. Sena, A.L. Martinez, Heat-Pipe Development for Advanced Energy Transport Concepts: Final Report Covering the Period January 1999 Through September 2001, Los Alamos National Laboratory, New York, NY, 2002 LA-13949-PR.

[43] C.L. Tien, A.R. Rohani, Theory of two-component heat pipes, J. Heat Transfer 94 (4) (1972) 479, https://doi.org/10.1115/1.344997.

[44] P. McCluew, et al., Design of Megawatt Power Level Heat Pipe Reactors, Las Alamos National Laboratory, New York, NY, 2015 LA-UR-15-28840.

[45] W.R. Determan, G. Hagelston, Thermionic in-core heat pipe design and performance, in: M.S. Elgenk, M.D. Hoover (Eds.), Proceedings of Ninth Symposium on Space Nuclear Power Systems, American Institute of Physics, CONF-920104, AIP-CP-2463, 1992, pp. 1046 https://doi.org/10.1063/1.41913.

[46] J. Zhang, et al., Redox potential control in molten salt systems for corrosion mitigation, Corros. Sci. 144 (2018) 44, https://doi.org/10.1016/j.corsci.2018.08.035.

[47] C.W. Forsberg, et al., Tritium control and capture in salt-cooled fission and fusion reactors: status, challenges, and path forward (critical review), Nucl. Technol. 197 (2017) 119, https://doi.org/10.13182/NT16-101.

[48] X. Wu, et al., Conceptual design of tritium removal facilities for FHRs, Presented at the 16th Int. Tool. Mtg. Nuclear Reactor Thermal Hydraulics (NURETH-16), American Nuclear Society, 2015 August 30–September 4.

[49] T. Tanabe, et al., Hydrogen transport in stainless steels, J. Nucl. Mater. 123 (1984) 1568, https://doi.org/10.1016/0022-3115(84)90304-0.

[50] A. Nakamura, S. Fukada, R. Nishiumi, Hydrogen isotopes permeation in a fluoride molten salt for nuclear fusion blanket, J. Plasma Fusion Res. 11 (2015) 25, https://pdfs.semanticscholar.org/4d8c/d2273bd3e3a 99129105ea7e1bc5d6edcaecb.pdf (Accessed 14 July 2019).

[51] R.A. Causey, R.A. Karnesky, C.S.A.N. Marchi, 4.16-Tritium barriers and tritium diffusion in fusion reactors, in: R.J.M. Konings (Ed.), Comprehensive Nuclear Materials, Elsevier, Oxford, 2012, pp. 511–549.

[52] J. Trouve, G. Laflanche, Hydrogen diffusion and permeation in sodium, Presented at the Int. Conf. on Liquid Metal Engineering and Technology in Energy Production, 1984 CEA-CONF-7645, April 9–13.

[53] J.M. Leimert, M. Dillig, J. Karl, Hydrogen inactivation of liquid metal heat pipes, Int. J. Heat Mass Transfer 92 (2016) 920, https://doi.org/10.1016/j.ijheatmasstransfer.2015.09.058.

[54] M.T. North and W.G. Anderson, "Hydrogen Permeation Resistant Heat Pipe For Bi-Modal Reactors: Final Report Period Covered," Thermacore, Oct 1–September 30, 1994 (1995); https://inis.iaea.org/collection/NCLCollectionStore/_Public/27/063/27063177.pdf (Accessed 14 July 2019).

[55] G. Parry, R.J. Pulham, Rate of reaction of hydrogen with liquid lithium: comparison with sodium and potassium, Dalton Trans. 19 (1975) 1915, https://doi.org/10.1039/dt9750001915.

[56] J.E. Hansel, et al., Sockeye: A 1-D Heat Pipe Modeling Tool, Idaho National Laboratory, New York, NY, 2019 INL/EXT-19-55742.

[57] J.E. Hansel, et al., Sockeye Theory Manual, Idaho National Laboratory, New York, NY, 2019 INL/EXT-19-54395.

[58] B. Zohuri, Functionality, Advancements and Industrial Applications of Heat Pipes, first ed., Academic Press, New York, NY, 2020.

[59] B. Zohuri, Nuclear Micro Reactors, first ed., Springer, New York, NY, 2020.

[60] B. Zohuri, Compact Heat Exchangers, Select, Application, Design and Evaluation, first ed., Springer, New York, NY, 2017.

[61] S.W. Chi, Heat Pipe Theory and Practice, McGraw-Hill, New York, NY, 1976.

CHAPTER 7

Salt cleanup and waste solidification for fission and fusion reactors

7.1 Introduction

From the 1950s through the early 1970s, there were large programs to develop molten salt reactors (MSRs) and the associated salt cleanup systems. This was followed by 40 years with little development work other than integral fast reactor chloride salt pyroprocessing separations for solid-fuel recycling. In the last 10 years, there has been a major revival in interest in three classes of salt-cooled reactors for different reasons. In each class of reactor, there are many design variants.

- *Fluoride-salt-cooled high-temperature reactor (FHR):* The FHR [1, 2] fuel is the graphite-matrix tri-structural isotropic (TRISO)-coated-particle fuel used in high-temperature gas-cooled reactors with a clean fluoride salt coolant and a nuclear air-Brayton combined cycle (NACC) [3-5]. The goal of this combination of technologies is to minimize reactor development risks (demonstrated high-temperature gas-cooled reactor fuel, clean fluoride salt coolants) to enable early deployment of a high-temperature reactor with NACC.

NACC can operate in multiple modes including base-load electricity production and peak power using stored heat or an external fuel (natural gas, oil, biofuels, hydrogen, etc.). The power cycle capability enables the reactor to (1) generate more revenue than a traditional base-load nuclear reactor with a steam cycle and (2) potentially replace fossil plants in their role of providing economic dispatchable electricity for a low-carbon world [6]. However, the power cycle requires heat inputs above ~600 °C that, in turn, dictates use of salt coolants.

- *Molten salt reactors:* In MSRs the fuel is dissolved in a fluoride or chloride salt. There is renewed interest in this class of reactors because of potential safety advantages and fuel cycle versatility (actinide burning to breeding). MSRs can also couple to NACC. The fluoride salt reactors can have a thermal [7, 8] to fast [9] neutron spectrum while the chloride salt MSRs [10-12] have fast neutron spectrums because of the high thermal cross section of chloride salts—particularly but not exclusively ^{35}Cl. The MSR programs in the 1950s and 1960s used fluoride salts with graphite or beryllium moderator to create thermal-spectrum reactors.

Much of the recent interest in molten chloride fast reactors (MCFRs) follows from the recognition that if isotopically separated chlorine-37 is used, it is possible to build a MSR with a very high breeding ratio and the possibility of a breed-and-burn fast reactor that (1) starts up on enriched uranium, spent nuclear fuel (SNF) or existing plutonium resources, (2) is refueled with natural uranium, (3) the salt with fissile material is treated as waste, and (4) ~30% of the uranium atoms as uranium or plutonium are fissioned in the once-through fuel cycle. There are also a variety of other fuel cycle options including cycles with very low levels of actinides to wastes. The cost of isotopically separating ^{35}Cl and ^{37}Cl create incentives to recover the ^{37}Cl for salts that become wastes.

- *Fusion with molten salt blankets*: Recent advances [13] in magnetic fusion systems have created large incentives for molten salt blankets using $^{6}Li_2BeF_6$ (flibe) that serve three functions: (1) generate tritium—the fuel of fusion machines, (2) recover the heat by slowing down the 14 MeV neutrons, and (3) radiation shielding. The development of new superconductors has enabled doubling of the magnetic fields in fusion systems. Fusion systems scale as one over the fourth power of the magnetic field; that is, doubling the magnetic field enables the fusion machine size to be reduced by an order of magnitude. However, this implies extremely high power densities where there are serious questions whether a solid fusion blanket can be cooled. The alternative design is to replace the solid fusion blanket by a tank of liquid salt.

7.2 Requirements

Traditional reactor coolants are simple: water, sodium, helium, and lead. The treatment process is to remove impurities and return clean coolant to the reactor. Salt reactor coolants are complex. All are mixtures of two or more salts to obtain acceptable physical properties such as acceptably low melting points. That creates more complex salt treatment processes (Fig. 7.1) to clean salts and meet all the requirements for salt to be returned to the reactor.

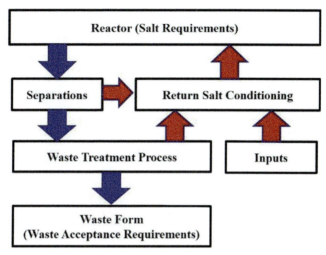

Fig. 7.1 Treatment of salts.

Salt-processing systems have two bounding sets of requirements:
1. Those defined by the reactor and
2. Those defined to meet the waste disposal requirements defined by the waste management system.

There are also a set of requirements on the separations systems. We describe some of the requirements and constraints before discussing technologies.

7.2.1 Reactor salt requirements

The choice of salt for each reactor type is based on the physical properties of the salt (melting point, viscosity, etc.) and nuclear properties. Because of the high neutron cross section of chlorine, only fluoride salts are candidates for fusion and thermal-spectrum fission reactors (FHR and fluoride MSRs).

Table 7.1 shows the major fluoride salts that are considered for FHRs and thermal-spectrum fluoride MSRs. The specific ratio of the major salt components may vary from what is shown. All candidate liquids are salt mixtures to lower melting points. For all salts containing lithium in fission reactors, isotopically separated ^{7}Li is used to avoid the high neutron adsorption cross section of ^{6}Li. The cost of ^{7}Li implies economic incentives to recycle the lithium.

For salt-cooled fusion reactors, the only salt considered is flibe (Li_2BeF_4) with lithium-6 because of the neutronic requirement to generate tritium (the fuel of a fusion reactor) as fast as it is consumed.

Table 7.1 Major fluoride salt coolant options.

Coolant	T_{melt} (°C)	T_{boil} (°C)	ρ (kg/m³)	PC_p (kJ/m³ °C)
66.7 ⁷LiF–33.3 BeF₂ (FLIBE	459	1430	1940	4670
59.5 NaF–40.5 ZrF₄	500	1290	3140	3670
26 ⁷LiF–37 NaF–37 ZrF₄	436		2790	3500
51 ⁷LiF–49 ZrF₄	509		3090	3750
Water (7.5 MPa)	0	290	732	4040

Compositions in mole percent. Salt properties at 700 °C and 1 atmosphere. Pressurized water data shown at 290 °C for comparison.

For the FHR and fusion reactors with clean salts, the salt processing requirements are driven by the need to minimize equipment corrosion, minimize radiation exposures to workers, and minimize tritium releases to the environment if the salt contains lithium or beryllium—elements that generate tritium upon neutron irradiation. Impurities are measured in 10s–100s of parts per million. A recent review paper [14] discusses the alternative options to capture tritium in the coolant to minimize release to the environment. All MSRs produce tertiary fission hydrogen, deuterium, and tritium but at much smaller rates if there is no lithium or beryllium in the salt.

For MSRs, there is the additional constraint that the fuel (uranium, plutonium) and fission products must be soluble in high concentrations in the salt. Other salts may be added to the base-line coolant for specific reasons. For example, most designs of fluoride MSRs contain some ZrF_4 to act as an oxygen getter in the event of oxygen ingress and remove the small amount of oxygen generated by interactions of neutrons with the coolant. The ZrF_4 reacts preferentially to uranium and plutonium fluorides to prevent forming insoluble uranium and plutonium fluorides that would precipitate from solution creating a potential criticality concern outside the reactor core or structural hot spots from local uncooled fissioning. Gas sparging is also used to remove oxygen and other impurities; thus, normally the ZrF_4 acts as a backup system. Other salts may be added for redox control [7, 9, 15] to minimize corrosion. In a fluoride MSR, the ratio of U^{+3}/U^{+4} is traditionally used for redox control.

There has been recent work on fluoride fast-spectrum MSRs in Europe. These reactors do not have neutron moderators in the reactor core. Most of the work has been using a salt consisting of $^7Li-ThF_4-UF_4-TruF_3$ (77.7–6.7–12.3–3.3 mol%) with typical uranium enrichments near 13%.

Historically, there has been limited work on MCFRs and thus the preferred salt compositions are not yet well defined. The relatively recent recognition that very high breeding ratios are possible including breed-and-burn once-through fast reactors has resulted in growth in interest in this type of MSR.

Many of the reported salt compositions [7] are eutectic mixtures. A typical composition is ~60% NaCl, 20% PuCl$_3$, and 20% UCl$_3$/lanthanide chlorides—all in mole percent. Other proposed mixtures include varying amounts of UCl$_4$. In most proposed systems the primary anion will be ^{37}Cl [12] to reduce neutron adsorption.

The sodium chloride is used to lower melting points [16]. Such a salt may contain minor eutectic components MgCl$_2$, KCl, SnCl$_4$, CaCl$_2$, ZrCl$_4$, and AlCl$_3$ to further reduce the melting point and viscosity or provide other functions. The mixture may include other cations for redox control. One will build in the fission product chlorides (CsCI, SrCl$_2$, etc.) as well as NaI (iodide FPs) and UI$_3$, which also tend to reduce the eutectic melting point and viscosity.

The important observation is that salt treatment (Fig. 7.1) is not just about removal of specific elements from the salt but controlling the chemical composition of a multicomponent mixture within some defined composition range.

7.2.2 Molten salt separations requirements

For the FHR and fusion reactors, the salt purification requirements are simple to define—a well-defined coolant composition. That is not true for MSRs where the salt composition will evolve over time because of fission and changing actinide compositions. There are a set of constraints that define allowable chemical composition: (1) maintain acceptably low salt melting points and viscosities, (2) avoid exceeding solubility limits of any actinide or fission product that would precipitate a selected element or elements out of solution, (3) maintain appropriate thermal expansion coefficient [negative temperature coefficient of reactivity], (4) maintain redox control to minimize corrosion, and (5) removal of fission products to avoid excessive parasitic neutron adsorption.

For some of these requirements the reactor can operate for years before a specific constraint becomes important; thus, there is the choice to remove selected fission products or actinides on-line or batch at the reactor or ship the salt off-site to a central facility for purification.

In one respect, MSR processing is different from recycle of solid fuels. In closed fuel cycles for solid-fuel reactors, fissile materials are recovered from SNF and fabricated into new nuclear fuel assemblies. To minimize radiation levels in the solid-fuel fabrication process, avoid problems such as volatile fission products during sintering and enable inspection of the fresh fuel assemblies, the traditional reprocessing plant goal has been to produce relatively pure fissile and fertile materials in specific proportions—a process that in many cases raises concerns about diversion of fissile materials (reducing proliferation resistance). In a MSR, "fresh fuel" is dumped into a molten salt mixture with large inventories of fission products and actinides. There is no incentive to create clean return coolant that may contain fuel; thus, most flow sheets have the goal to remove fission products with relatively low separations efficiency, or not at all, to minimize costs.

Fuel cycle goals may drive MSR separations requirements. Thermal-spectrum fluoride MSRs have low fissile inventories; but, they must limit fission product inventories in the reactor core to limit parasitic neutron adsorption. Fast-spectrum MSRs [9] have larger fissile inventories and can be breeder reactors with fewer constraints on allowable fission product inventories because of the lower absorption cross section of fission products.

Last, for some MCFRs it is proposed to have a once-through fuel cycle with no separations (except noble gases and noble fission products) to minimize (1) costs and (2) risks of proliferation by maximizing fission products associated with plutonium. After startup of the MCFR, it is a natural uranium salt mixture in with an equivalent quantity of salt with uranium, plutonium, actinides, fission products, and salt components to waste. Neutronically, such a fuel cycle that converts more than 30% of the feed uranium to fission products is viable.

It is unclear today whether some separations will be required to meet the various long-term MCFR requirements. However, in either case it implies that there may be salt reactors where economics or nonproliferation goals generate wastes with significant quantities of uranium and plutonium and require waste forms with significant loadings of uranium and plutonium.

7.2.3 Final waste form requirements

Final waste form choices will depend upon:
1. the chemical and radioisotopic content of the waste,
2. legal requirements, and
3. waste disposal site acceptance criteria.

The requirements for low-level wastes as might be generated by an FHR or a salt-cooled fusion machined are different than a MCFR waste with a high concentration of fission products and actinides that requires geological disposal. We discuss herein waste forms suitable for disposal of any high-level waste (HLW) —high-performance HLW forms. Depending upon economics, these waste forms may be used for less hazardous halide wastes.

There has been a great deal of work on developing high-quality HLW forms in the last 40 years. Processes to convert aqueous nitrate HLWs into final HLW forms have been developed and deployed on an industrial scale. Borosilicate HLW glass is produced in the United States, France, and United Kingdom. Russia produces an iron phosphate HLW form. The choice of waste form depends upon the waste to be solidified. Borosilicate glasses were developed for wastes containing significant quantities of sodium; thus, may become the preferred form for MCFRs where there is significant sodium in the waste stream; however, no studies have been done assessing the option space.

There have been no significant efforts to develop processes to convert chloride or fluoride salt reactor wastes into high-quality HLW forms. It is our perspective that the likely HLW forms are borosilicate and iron phosphates for multiple reasons:

1. These are today the only two generally accepted high-quality HLW forms.
2. The costs of developing and qualifying new HLW forms in dollars and time is large.
3. There are several methods to convert halide salts to these waste forms.
4. It is difficult to develop high-quality halide salt waste forms with high waste loadings.

Waste forms with low waste loadings imply high storage, transport, and disposal costs.

There are several reasons why chloride HLW forms are unlikely. First, most chlorides have significant solubility in water and therefore are poor waste forms. The high solubility of most chlorides is why oceans are filled with salt water and the continents are made of oxides. Second, most MCFR designs use isotopically separated ^{37}Cl that creates a strong incentive to recycle the ^{37}Cl back to the reactor to avoid the cost of added chlorine enrichment. Third, if ^{35}Cl (the other isotope of chloride) is irradiated in a reactor, it generates long-lived ^{36}Cl that in repository assessments [17] is one of the few radionuclides that can leak from a geological repository. This creates separate waste management incentives to use relatively pure ^{37}Cl in chloride salt reactors and added incentives to recycle chlorides from MCFRs back to the reactor. Work is underway to convert chloride wastes from pyroprocesses used for metallic fuel recycle into chloride waste forms for disposal; however, these are normal chlorides that have not been irradiated (no ^{36}Cl) and are expected to have low concentrations of long-lived actinides.

The challenge for many fluoride waste forms in high radiation fields is radiolysis [18-19]. Many frozen fluoride salts at low temperatures will generate free fluorine or gaseous fluoride compounds. In the Molten Salt Reactor Experiment, this free fluorine combined with uranium to form volatile UF6. It is not well understood how general this radiolysis pathway is in fluorides. If radiolysis is a concern in the final waste form, the options are to design a waste form with a chemical getter to control free fluorine or separate the fluorine from the waste stream.

7.3 Separations

There are multiple separations options [7] to remove corrosion products and fission products from salts creating (1) a salt stream that is returned to the reactor and (2) a salt stream that is a waste for conversion into a high-quality waste form. In MSRs the salt stream that is returned may include added uranium and other salt components to produce the desired "fresh" salt—although that stream may also contain large quantities of fission products and actinides.

We discuss herein only separations associated with soluble chlorides or halides. During operation of a MSR noble fission product gases (Xe and Kr) and HF/HCl exit via the off-gas system. Noble metal fission products form multiatom clusters in the salt that plate out on surfaces. These insoluble fission products can be removed by gas sparging or high-surface area beds [14] and thus reduce noble metal plate out on heat exchanger surfaces. These are unavoidable separations that occur in the reactor. The rest of the fission products form soluble chlorides or fluorides.

Reductive extraction into liquid bismuth and fluorination of uranium to UF6 were first developed for salt processing in the 1970s. We discuss herein the "new" options made available since the 1970s. Different separations processes operate on different mechanisms; thus, a separation that may be difficult with one process may be easy using a different process. For example, in chloride systems separation of CsCl using electrochemical processes is generally difficult because cesium salts usually have a more negative redox potential than the bulk salt and actinides (bulk salt electrolyzes first) while it appears easy to separate CsCl by distillation because it has a lower boiling temperature than NaCl.

7.3.1 Distillation

Distillation is the dominant separation process in the chemical industry because of simplicity and no need to add a chemical reagent. It was experimentally examined during the early development of fluoride MSRs [20-22] but not adopted—partly because of the difficulty to fabricate refractory high-temperature distillation equipment made of molybdenum or other high-temperature materials. Several decades of advances in fabrication technology have eliminated that barrier making very high-temperature distillation a credible option.

The Chinese [8] are developing a single stage vacuum distillation process for separating fission products for their thermal-spectrum fluoride MSR that will operate on a ^{232}U–thorium fuel cycle. The current Chinese salt-processing flow sheet for a ^{233}U–thorium fluoride MSR involves:
1. removal of the uranium by fluorination,
2. distillation that recovers most of the valuable flibe salt,
3. return of salt and uranium to the reactor, and
4. off-site chemical separation of thorium and residual fissile materials from the lower-value salt residue that contains most of the fission products.

This flow sheet enables economics of scale for the more complex separations off-site with rapid recycle of the valuable fissile materials and salt at the reactor.

Distillation for purifying FHR and salt-cooled fusion salts will be significantly different than for MSRs. The FHR and salt-cooled fusion impurities will include corrosion products, various other impurities that entered the system by accident and for the FHR fission products from failed fuel. For FHR and fusion, the goals are well defined—a clean high-purity salt. The feed impurity levels will be measured in at most a few hundred parts per million implying relatively clean feeds with distillation columns requiring multiple stages with high-efficiency packings to provide pure salts. In the last 5 years additive manufacturing technologies have been developed to fabricate high-temperature three-dimensional components such as turbine blades that could be used to fabricate such high-temperature distillation columns with high-performance packings. Depending upon goals, one or more distillation columns may be required.

Distillation for MSRs implies salts with high fission product loadings, high actinide loadings that imply criticality limits that will limit the diameter of the distillation column and significant decay heat. It may or may not require multistage distillation systems. There are complications.
- *Noble metals will plate out in the distillation system.* This may be good or bad for there is the option to use replaceable distillation columns and use the distillation column as the waste package for the noble metals.
- *Uranium and many of the actinides have multiple valence states, thus the redox must be controlled to determine behavior in the column.* For example, uranium

in the UCl_3 form has a boiling point of 1657 °C, whereas UCl_4 has a boiling point of 791 °C. Valence states can change with temperature and there may be compounds that thermally decompose at higher temperatures.

- *Internal heat generation will change distillation column design.* Distilling molten salts with high fission product loadings implies high decay heat generation—requiring careful distillation column design.

There are some distillation columns in the chemical industry, where chemical reactions that generate significant heat are initiated by catalysts—the chemical reactor and the separation system are combined. This experience provides some basis for the design of such systems.

- *Incentives for higher pressure distillation.* Increasing pressure (avoiding vacuum distillation) reduces equipment sizes—but raises temperatures. Improved fabrication methods may enable higher temperatures.

In the chloride system (Table 7.2), there are several different ways distillation may be used depending upon goals. For example, UCl_3 has a boiling point of 1657 °C that is close to $PuCl_3$ with a boiling point of 1767 °C, so the uranium and plutonium remain together. The option may exist to allow distillation to remove the lower boiling point chlorides including CsCl (1297 °C), $SrCl_2$ (1250 °C), and many of the rare earth chlorides ($LaCl_3$, 1000 °C).

Processes such as distillation will feed back into reactor design and may require rethinking of design goals. Americium chloride ($AmCl_3$) has a low boiling point (850 °C)—implying that in a distillation system it will be part of the low-boiling fraction while other fissile and fertile materials will be in the high-boiling fraction. This is good or bad depending

Table 7.2 Melting and boiling points of selected chlorides.

Compound	Melting point (°C)	Boiling point (°C)
$ZrCl_4$	437	331
CsCl	645	1297
$SrCl_2$	874	1250
$MgCl_2$	714	1400
NaCl	801	1413
KCl	770	1420
UCl_3	837	1657
UCl_4	590	791
$LaCl_3$	858	1000
$PuCl_3$	780	1767
$AmCl_3$	750	850

upon ones perspective. If it is removed, higher actinide generation will be less. However, it is also the major generator of decay heat in a repository 60–70 years after reactor discharge and decay of ^{137}C and ^{90}Sr.

Fig. 7.2 [23] shows some of the measured vapor pressure curves of different chlorides. However, significant work remains to develop distillation for many of these separations. Most of the work on chloride distillation has been associated with the lower-temperature purification of zirconium—both from ores (Kroll process) and for recycle of LWR SNF clad. There has also been some work done on chloride-based reprocessing flow sheets—again because of the ability of volatile chloride processes

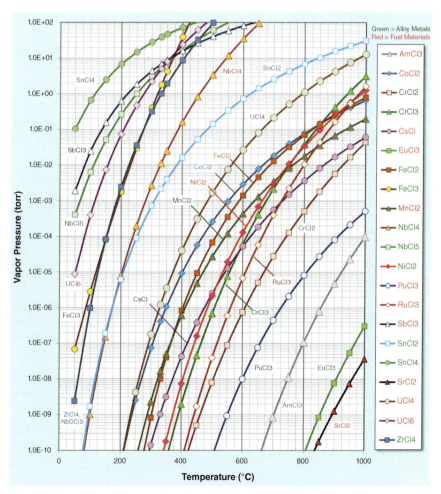

Fig. 7.2 Vapor pressure curves for selected chlorides [Ref. 23].

to separate zirconium from fuels with significant zirconium in the fuel element. However, there are several caveats. The species in solution can have very different volatility due to complexation. One must consider all radionuclides—including ones that are important in the context of many repositories such as iodine and technetium. Much experimental work is required before reliable predictions can be made of performance.

7.3.2 Electrochemical separations

Electrochemical separation processes [24, 25] have been developed for metallic sodium fast reactor fuel. In simplified terms:
1. The fuel is chopped into pieces and put into a bath of lithium and potassium chloride.
2. Noble metals drop to the bottom of the basket.
3. Uranium and plutonium are transferred to the other electrode by applying an electric current.
4. The remaining fission products form stable chlorides.

Variants of this process may be viable to purify chloride MSR salts with recovery of uranium and actinides followed by conversion back to chlorides and mixing with fresh base salt before recycling to the reactor. For MCFRs, the process would be somewhat simplified—no solid fuel and no requirement for high separation efficiencies. However, there are complications because the lithium–potassium salt used for sodium fast reactor pyroprocessing was chosen for efficient separations, whereas in a MCFR the salt is chosen to match reactor requirements. One does not get to choose the base salt composition—it is chosen based on reactor requirements, not separation goals.

- *Sodium chloride waste stream.* In such an electrochemical process the base salt (mostly sodium chloride) with the fission products is the waste stream. After removal of uranium and plutonium, one may have a sodium chloride salt with a relatively dilute concentration of fission products implying low fission product loadings in the final waste form. If one pushes for greater fission product removal (higher processing rate), the ratio of fission product chlorides to sodium chloride decreases with increased final waste volumes.
- *Limited electrochemical separations.* Depending upon the redox condition chosen for the MCFR, there may be some fission product chlorides less stable than the actinide chlorides and other fission product chlorides more stable than the actinide chlorides. In this case, the electrochemical cell will first plate out the less stable fission product chlorides. If

the goal is removal of most fission product chlorides, one may have an electrolytic cell for first removing less stable fission product chlorides and a second cell to remove actinides. Alternatively, one could use the electrochemical process to only remove the less chemically stable fission product chlorides with recycle of the salt directly back to the reactor.

For FHR and salt-cooled fusion reactors, there is the option of electrochemical cleanup of the salt—but only limited work has been done to date. The impurities in these systems tend to be the metals of construction that will plate out as metals at lower voltages than the voltage that decomposes the clean salt.

7.4 Conversion of salt wastes to high-quality waste forms

There have been no large-scale development programs for salt reactor wastes; however, many wastes contain halides and thus there has been work on converting halide wastes into qualified waste forms. Much of the work on processes for conversion of halide wastes into high-quality waste forms were developed for other purposes—such as treatment of plutonium residues that contain some quantity of chlorides or fluorides. In each case a compound is added (H_3PO_4, HNO_3, PbO, etc.) that reacts with the chloride or fluoride to form an oxide and a volatile chloride (HCl, HF, $PbCl_2$, etc.) that goes into the off-gas, where the halide can be recovered for recycle—most likely by scrubbing with sodium hydroxide or potassium hydroxide to produce salts that can be recycled.

The choice of process will be determined by economics when considering the process (complexity, ease of operations, secondary waste generation, etc.) and the final waste form. The allowable waste loadings in phosphate and borosilicate waste forms depend upon the specific composition of the waste. If one waste form has several times the waste loadings of the other waste form, waste form may determine process choices.

7.4.1 Conversion of halide wastes to iron phosphate gas

Chloride salts can be converted to iron phosphate waste forms [26] by addition of phosphoric acid and ferric (iron) oxide to the salt melt. The chloride reacts with phosphoric acid and is converted to an oxide with HCl going to the off-gas system. In the off-gas system the chloride can be recovered by scrubbing the off-gas with an aqueous hydroxide. For the MCFR, the scrubber would use sodium hydroxide to enable recovery of the ^{37}Cl for recycle to the reactor as NaCl. With the addition of ferric oxide, the final product is an iron phosphate glass—a generally accepted HLW form.

Fluoride salts can be converted to iron phosphate waste forms [27] by a similar process. The fluoride salt reacts with nitric acid and is converted to a nitrate with HF going to the off-gas system. In the off-gas system, the halide can be recovered by scrubbing the off-gas with an aqueous hydroxide. The nitrate is then fed to a glass melter with additives to produce an iron phosphate glass—a generally accepted HLW form. Nitric acid is a stronger acid and thus better at converting metal fluorides to oxides and HF readily boils off. The off-gas includes NO_x that can be catalytically decomposed with no residuals.

The chemistry for all the process steps has been demonstrated but there has been no process engineering. No detailed assessment of the option space has been done—including the chemical engineering and testing of materials in terms of corrosion in these environments.

7.4.2 Conversion of halide wastes to borosilicate glass

One waste processing flow sheet has been demonstrated at a proof-of-principle laboratory scale for conversion of salt wastes to borosilicate glass [28, 29]. Borosilicate glass was originally developed for converting a high-sodium HLW at the Savannah River Site into a repository-acceptable sodium borosilicate glass. The salt is mixed with a mixture of boron oxide and lead oxide—a fusion melt designed to dissolve highly refractory compounds. The lead oxide reacts with chlorides to form oxides and volatile lead chloride that goes to the off-gas system. The resultant boron fusion melt can be mixed with silicon oxide and other cations to form a borosilicate glass. The scrubber in the off-gas can recover the chloride as a sodium or potassium chloride (see Fig. 7.3).

A similar set of reactions occur with fluoride wastes. There are other possible oxidants such as nitric acid for conversion from a halide to an oxide system; but, the author is unaware of whether all the process steps to borosilicate glass have been demonstrated in the laboratory.

7.5 Other considerations

With salt reactors there is the option to conduct separations and waste treatment at the reactor site or off site at a central facility or some combination. The preferred option or options will depend upon economics. If there are large sites with multiple reactors, it may favor on-site processing

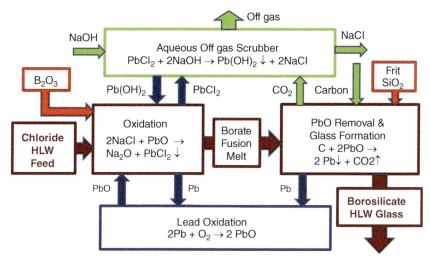

Fig. 7.3 Converting halide wastes to borosilicate HLW glass.

whereas reactor sites with small output will likely favor centralized processing because of economics of scale.

It has been recognized that in MSRs on-line processing of the salt creates the option to minimize the inventory of longer-lived fission product radionuclides in the reactor and thus minimize:
1. radioactive decay heat,
2. the accident source term, and
3. the potential for land contamination.

For that option to be real, simplified methods to separate selected radionuclides and solidify them are required. Chemical separations are not sufficient—it just moves the radioactivity from one pot to another pot within the plant. One must be able to separate selected radionuclides, convert them into a high-integrity waste form and store that waste form in a system with passive cooling such as dry-cask storage. The combination of improved separation (distillation) and solidification processes may create such an option. Whether such an option is worthwhile or even credible involves tradeoffs between accident risks and added risks associated with salt processing—along with economic considerations.

7.6 Conclusions

Much work was done in the 1950 and 1960s on fluoride salt processing for MSRs, but there was no significant work in how to convert salt wastes into repository-acceptable HLW forms. Since then there has been massive work on development of HLW solidification processes, separation processes, and process equipment for other wastes—some that included halides.

These developments create multiple options that did not exist several decades ago. What is now required is to develop flow sheets combined with experimental work to provide the underlying physical property data bases. From this starting point, the various options can be evaluated to potentially develop low-cost back-end fuel cycles that meet the different requirements for different salt reactor systems. Significant added work is required to sort out realistic options based on the currently available technologies.

References

[1] C. Andreades, et al., Design summary of the mark-I pebble-bed, fluoride-salt-cooled, high-temperature reactor commercial power plant, Nucl. Technol. 195 (2016) 223–238.

[2] C. Forsberg, P.F. Peterson, Basis for fluoride-salt-cooled high-temperature reactors with nuclear air-Brayton combined cycles and firebrick resistance-heated energy storage, Nucl. Technol. 196 (2016), http://dx.doi.org/10.13182/NT16-28.

[3] C. Andreades, R.O. Scarlat, L. Dempsey, P.F. Peterson, Reheating air-Brayton combined cycle power conversion design and performance under normal ambient conditions, J. Eng. Gas Turbines Power 136 (2014).

[4] C. Forsberg, P.F. Peterson, P. McDaniel, H. Bindra, Nuclear combined cycle gas turbines for variable electricity and heat using firebrick heat storage and low-carbon fuels, Proceedings of ICAPP 2017, Paper 17567, Fukui and Kyoto (Japan), 2017, April 24–28.

[5] C. Forsberg, et al., Converting excess low-priced electricity into high-temperature stored heat for industry and high-value electricity, Electr. J. (2017), https://doi.org/10.1016/j.tej.2017.06.009.

[6] C.W. Forsberg, et al., MIT-Japan Study: Future of Nuclear Power in a Low-Carbon World: The Need for Dispatchable Energy, Center for Advanced Nuclear Energy (CANES), Massachusetts Institute of Technology, 2017. MIT-ANP-TR-171, http://energy.mit.edu/wp-content/uploads/2017/12/MIT-Japan-Study-Future-of-Nuclear-Power-in-a-Low-Carbon-World-The-Need-for-Dispatchable-Energy.pdf.

[7] T.J. Dolan (Ed.), Molten Salt Reactors and Thorium Technology, Woodhead Publishing, Oak Ridge Tennesi, USA, 2017.

[8] H. XU, China's TMSR programme. In: Workshop on Molten Salt Technologies, Oak Ridge National Laboratory, 2015, https://public.ornl.gov/conferences/msr2015/Presentations.cfm.

[9] EUROPEAN UNION, *Safety Assessment of the Molten Salt Fast Reactor*, http://samofar.eu/ or http://samofar.eu/summerschool/presentations-msr-summer-school-2-4-july-2017/.

[10] J.R. Cheatham III, et al., Fission Reaction Control in a Molten Salt Reactor, U.S. Patent: US2016/0189806 A1, 2016.

[11] A.T. Cisneros Jr, et al., Molten Nuclear Fuel Salts and Related Systems and Methods, U.S. Patent: US2016/0189813 A1, 2016.

[12] B. Hombouger et al., "Fuel cycle analysis of a molten salt reactor for breed-and-burn mode," Paper 15524, *Proceedings ICAPP 2015*, Nice, France.

[13] B.N. Sorbom, et al., ARC: a compact, high-field, fusion nuclear science facility and demonstration power plant with demountable magnets, Fus. Eng. Des. 100 (2015) 378–405, http://dx.doi.org/10.1016/j.fusengdes.2015.07.008, ISSN 0920-3796.

[14] C.W. Forsberg, S. Lam, D.M. Carpenter, D.G. Whyte, R. Scarlat, C. Contescu, L. Wie, J. Stempien, E. Blandford, Tritium control and capture in salt-cooled fission and fusion reactors: status, challenges, and path forward (critical review), Nucl. Technol. 197 (2017), https://doi.org/10.13182/NT16-101.

[15] S.T. Lam, R. Ballinger, C. Forsberg, Controlling corrosion and tritium in a fluoride-salt-cooled high-temperature reactor using hydrogen, Transactions American Nuclear Society Winter Meeting, 2017.

[16] E.S. Sooby, et al., Measurements of the liquidus surface and solidus transitions of the $NaCl–UCl_3$ and $NaCl–UCl_3–CeCl_3$ phase diagrams, J. Nucl. Mater. 466 (2015) 280–285.

[17] B. Faybishenko, J. Birkholzer, D. Sannani, P. Swift, International Approaches for Nuclear Waste Disposal in Geological Formations: Geological Challenges in Radioactive Waste Isolation—Fifth Worldwide Review, LBNL, Oak Ridge Tennesi, USA, 1006984, 2017.

[18] L.M. Toth, L.K. Felker, Fluorine generation by gamma radiolysis of a fluoride salt mixture, Radiat. Effects Defects Solids 112 (4) (1990) 201–210, https://doi.org/10.1080/10420159008213046.

[19] L.D. Trowbridge, A.S. Icenhour, D.G. Del Cul, D.W. Simmons, Radiolytic processes during intermediate states of ^{233}U removal from the Oak Ridge Molten Salt Reactor Experiment. In: Fifth Topical DOE Spent Nuclear Fuel and Fissile Materials Management, Charleston, South Carolina, 2002.

[20] J.R. Hightower Jr., L.E. McNeese, Low-Pressure Distillation of Molten Fluoride Mixtures: Nonradioactive Tests for the MSRE Distillation Experiment, ORNL-4434, (1971).

[21] J.R. Hightower Jr., L.E. McNeese, Measurement of the Relative Volatilities of Fluorides of Ce, La, Pr, Nd, Sm, Eu, Ba, Sr, Y. and Zr in Mixtures of LiF and BeF_2, ORNL-TM-2058, (1968).

[22] F.J. Smith, L.M. Ferris, C.T. Thompson, Liquid–Vapor Equilibria in $LiF–BeF_2$ and $LiF–BeF_2–ThF_4$ Systems, ORNL-TM-4415, (1969).

[23] D.F. McLaughlin, T.L. Francis, Recycle of Zirconium From Used Nuclear Fuel Cladding: Chlorination Process Development and Design, IDIQ Subtask 2, Westinghouse Report EPT-WZ-14-004, Rev. 1, (2014).

[24] W. Zhou, Y. Wang, J. Zhang, Integrated model development for safeguarding pyroprocessing facility: part I—model development and validation, Ann. Nucl. Energy 112 (2018) 603–614.

[25] M.A. Williamson, J.L. Willit, Preprocessing flowsheets for recycling used nuclear fuel, Nucl. Eng. Technol. 43 (2011) 329–334.

[26] D.D. Siemer, Improving the integral fast reactor's proposed salt waste management system, Nucl. Technol. 178 (2012) 341–352.

[27] D.D. Siemer, Molten salt breeder reactor waste management, Nucl. Technol. 185 (2013) 1–9.

[28] C.W. Forsberg, Treatment of Halogen-Containing Waste and Other Waste Materials, U.S. Patent: 5,613,241 (1997).

[29] C.W. Forsberg, et al., Direct Vitrification of Plutonium-Containing Materials (PCMs) With the Glass Material Oxidation and Dissolution System, Oak Ridge National Laboratory, ORNL-6825, (1995).

APPENDIX A

A combined cycle power conversion system for small modular LMFBR

A.1 Introduction

In previous publications, we have addressed applying an open-air-Brayton power conversion cycle to next generation nuclear power plants [1,2,3]. Both a combined cycle (CC) power conversion system and a recuperated Brayton cycle look very promising at the temperatures anticipated for molten salt reactors and liquid lead cooled reactors. However, at coolant temperatures more typical of a liquid metal fast breeder reactors (LMFBRs), the CC did not seem to hold an advantage over current conversion systems giving cycle efficiencies in the 39%–40% range. Recently taking a clue from standard steam cycle efficiency enhancement techniques, we added a "feed water heater" to the Rankine bottoming cycle [6]. This is not the classic feed water heater because it draws its heat by cooling the compressed air from the first stage of a split compressor in the Brayton topping cycle. This frees up most of the hot air that would be used to raise the water to the saturation temperature for the Rankine cycle so that it can be passed through a recuperator to preheat air prior to entering the main heat exchangers. The net effect of the process is to raise the thermal efficiency of the cycle about 3% making a CC more competitive at LMFBR outlet temperature.

A.2 The air-Brayton cycle, pros and cons

The fastest growing power conversion systems in the electric utility market today are internal combustion gas turbines. Any external combustion or heat engine system is always at a disadvantage to an internal combustion system. The internal combustion systems used in current jet engine and gas turbine power systems can operate at very high temperatures in the fluid, and cool the structures containing the fluid to achieve high thermodynamic efficiencies. In an external energy generation system, like a reactor, all of the components from the core to the heat exchangers heating the working fluid must operate at a higher temperature than the fluid.

This severely limits the peak cycle temperature compared to an internal combustion system. One way this liability can be overcome by using multiple expansion turbines and designing highly efficient heat exchangers to heat the working fluid between expansion processes.

Typically, the combustion chamber in a gas turbine involves a pressure drop of 3%–5% of the total pressure. Efficient liquid metal to air heat exchangers can theoretically be designed with pressure drops of less than 1%. This allows three to five expansion cycles to achieve a pressure drop comparable to a combustion system. Multiple turbines operating at different pressures have been common in steam power plants for a number of years. In this study three to five gas turbines operating on a common shaft were considered. Multiple expansion turbines allow a larger fraction of the heat input to be provided near the peak temperature of the cycle. The exhaust from the last gas turbine still contains significant amounts of thermal energy.

This gas is provided to the heat recovery steam generator to produce the steam used in the Rankine bottoming cycle. In a traditional gas turbine CC this gas gives up its thermal energy by heating steam in a superheater, then vaporizing the high-pressure water to make steam, and finally heats the high-pressure water to the liquid saturation point prior to evaporating it.

At the liquid saturation point or "pinch point" the hot air must be at a temperature greater than the boiling temperature of the high-pressure water. The heat remaining in the gas is then used to heat the liquid water or exhausted to the atmosphere.

A.3 The feed water heater

Traditional Rankine steam cycles bleed steam from the high pressure and temperature steam in the turbine(s) to heat the high-pressure water from low temperature to its boiling point. Because the temperature drop between the heating fluid and heated fluid in this case is less than in the case where the liquid water is heated by the boiler directly, the overall efficiency of the cycle increases.

For the CC, there is another source of heat that can be used to heat this low-temperature, high-pressure water. If the air compressor in the Brayton cycle is split into two parts such that the first part does about 56% of the work, its air can be used to heat the low-temperature water. The heat exchanger that cools the air exiting the first compressor is normally called an intercooler.

Traditionally, an outside source of cold water is used to perform this function. However, the cold water in the Rankine bottoming cycle can perform this function almost as well. This frees up the hot gas exiting the "pinch point" to preheat the air out of the second compressor in a traditional recuperator. Intercooling without recuperating does not normally improve the efficiency of a cycle [4,5].

The major limitation on the size of the steam system is the enthalpy available from high-temperature air above the pinch point where the high-pressure water working fluid starts to vaporize. Below this point, there is still a significant enthalpy in the air which is readily available to heat the high-pressure water or to heat the compressed air in the Brayton cycle. There does not appear to be an advantage to including traditional feed water heaters in the cycle to bring the high-pressure water up to the saturation point.

The possibility that an intercooler could be inserted between the two stages of a split compressor was considered. The cooling fluid for the intercooler was the high-pressure water coming out of the water pump. This process would combine the function of the traditional intercooler with the preheating of a typical feed water heater.

A.4 Results

The results of modeling this process for a temperature typical of an LMFBR (510°C or 783 K) are presented in Fig. A.1.

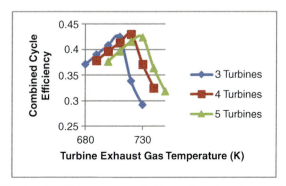

Fig. A.1 Combined cycle efficiency versus turbine exhaust gas temperature.

Table A.1 Comparison of air and steam pressure.

	Compression ratio	Steam pressure (MPa)
Three turbines	3.846	21.0
Four turbines	4.642	15
Five turbines	5.000	13.5

Basically, the addition of the intercooler-feed water heater increased the efficiency of the cycle by about 3%. The peak efficiency achieved by the four turbine case is 42.95%. The four turbine case is once again optimal as it was for the simple CC, though the margin over three or five turbines is less than a per cent. There were significant differences from the simple CC, however. Because the Brayton cycle is now recuperated, the overall compression ratios are much smaller. Also, the operating pressures in the Rankine cycle were much higher fed by the need to provide hot gas from the "pinch point" to the recuperator (see Table A.1).

A more meaningful comparison can be made with the four turbine case presented previously for the Molten Salt reactor. Simple CC is the previous case, intercooled combined cycle is the split compressor in Table A.2.

The cycle optimizes with the same turbine exit temperature, but a significantly different compression ratio. It is worth pointing out that with the addition of an intercooler-feed water heater, the molten salt efficiency is now very close to 50%.

The cycle optimizes with the same turbine exit temperature, but a significantly different compression ratio. It is worth pointing out that with the addition of an intercooler-feed water heater, the molten salt efficiency is now very close to 50%.

Returning to the LMFBR case, for the optimum four turbine configuration the major performance parameters for a 50 MW (electric) system would be (see Table A.3).

Note once again that the heat rejection to a circulating water system is significantly less than the heat rejected by a current coal fired plant at 40% efficiency. It is approximately 39.2% of that required by this kind of steam plant.

Table A.2 Comparison of intercooled with simple combined cycle.

	Exhaust temp (K)	Compress ratio	Steam pressure (MPa)	Overall efficiency
Simple CC	820	14.125	2	46.1
IC CC	820	10.488	15	49.5

Table A.3 Performance parameters for 50 MW(e) system.

Total power	50 MW(e)
Brayton system	31.2 MW(e)
Rankine system	18.8 MW(e)
Total heat input	116.0 MW(t)
Mass flow rate (air)	291.4 kg/s
Mass flow rate (water)	12.65 kg/s
Heat rejection	29.4 MW(t)

Perhaps even more interesting are the state points for the CC system. Table A.4 gives the typical state points for both the air-Brayton cycle and the steam Rankine cycle.

It is probably worth mentioning that the intercooler-feed water heater could probably be adapted to current generation gas turbine systems if the

Table A.4 Typical state points for both the air Brayton cycle and the steam Rankine cycle.

State points	Pressure (MPa)	Temperature (K)
Brayton		
Atmosphere	0.1013	288.0
First compressor exit	0.2489	382.9
Intercooler exit	0.2464	319.9
First heater inlet	0.4610	604.6
First heater exit	0.4563	783.0
First turbine exit	0.3204	720.0
Fourth turbine exit	0.1076	720.0
Superheater exit	0.1066	664.1
Evaporator exit	0.1053	625.3
Recuperator inlet	0.1044	625.3
Recuperator exit	0.1034	402.4
Exhaust to atmosphere	0.1023	402.4
Rankine		
Pump entrance	0.0073	313.0
Pump exit	15.0	316.6
Intercooler exit	15.0	615.3
Evaporator exit	15.0	615.3
HP turbine inlet	15.0	705.0
HP turbine exit	3.75	519.7
MP turbine inlet	3.75	705.0
MP turbine exit	0.9375	519.7
LP turbine inlet	0.9375	705.0
LP turbine exit quality	0.914	

low-temperature gas exiting the recuperator does not condense the water added to the air as a result of combustion. That question is beyond the scope of our effort.

References

[1] P. McDaniel, C. De Oliviera, B. Zohuri, J. Cole, A combined cycle power conversion system for the next generation nuclear plant, ANS Trans. (2012).
[2] B. Zohuri, P. McDaniel, C. De Oliviera, A comparison of a recuperated open cycle (air) Brayton power conversion system with the traditional steam Rankine cycle for the next generation nuclear power plant, ANS Trans. (2014).
[3] B. Zohuri, P. McDaniel, C. De Oliviera, Advanced nuclear open air-Brayton cycles for highly efficient power conversion, Nucl. Technol.submitted for publication, (2014).
[4] D.G. Wilson, T. Korakianitis, The Design of High-Efficiency Turbomachinery and Gas Turbines, second ed., Prentice-Hall, Upper Saddle River, NJ, 1998.
[5] R.W. Haywood, Analysis of Engineering Cycles, fourth ed., Pergamon Press, Oxford, 1991.
[6] M.M. El-Wakil, Powerplant Technology, McGraw-Hill, New York, NY, 1984.

APPENDIX B
Direct reactor auxiliary cooling system (DRACS)

B.1 Introduction

The direct reactor auxiliary cooling system (DRACS) has been proposed for advance high temperature reactor (AHTR) as the passive decay heat removal system. The DRACS features three coupled natural circulation/convection loops relying completely on buoyancy as the driving force. In the DRACS, two heat exchangers, namely, the DRACS heat exchanger (DHX) and the natural draft heat exchanger (NDHX) are used to couple these natural circulation/convection loops. In addition, a fluidic diode is employed to restrict parasitic. In addition, a fluidic diode is employed to restrict parasitic flow during normal operation of the reactor and to activate the DRACS in accidents.

While the DRACS concept has been proposed, there are no actual prototypic DRACS systems for AHTRs built and tested in the literature. In this report, a detailed modular design of the DRACS for a 20-MWth fluoride-salt cooled high temperature reactor (FHR) is first developed. As a starting point, the DRACS is designed to remove 1% of the nominal power, that is, the decay power being 200 kW.

The design process for the prototypic DRACS involves selection of the salts, identification of the reactor core, design of the DHX and NDHX, design of the fluidic diode, design of the air chimney, selection of the loop pipes, and finally determination of the loop height based on pressure drop analysis. FLiBe with high enrichment in Li-7 and FLiNaK have been selected as the primary and secondary salts, respectively. A 16-MWth pebble bed core proposed by University of California at Berkeley is adopted in the design. Shell-and-tube heat exchangers have been designed based on Delaware Method for the DHX and NDHX.

A vortex diode that has been tested with water in the literature is adopted in the present design. Finally, pipes with inner diameter of 15 cm are selected for both the primary and secondary loops of the DRACS. The final DRACS design features a total height less than 13 m. The design presented here has the potential to be used in the planned small-scale FHR test reactor and will also benefit and guide the DRACS design for a commercial AHTR.

Following the prototypic DRACS design is the detailed scaling analysis for the DRACS, which will provide guidance for the design of scaled-down DRACS test facilities. Based on the Boussinesq approximation used in the field of buoyancy-driven flow (i.e., also known as natural convection) in one-dimensional formulation, the governing equations, that is, the continuity, integral momentum, and energy equations are nondimensionalized by introducing appropriate dimensionless parameters, including the dimensionless length, temperature, velocity, etc. The key dimensionless numbers, that is, the Richardson, friction, Stanton, time ratio, Biot, and heat source numbers that characterize the DRACS system, are obtained from the nondimensional governing equations (see Fig. B.1).

Based on the dimensionless numbers and nondimensional governing equations, similarity laws are proposed. In addition, a scaling methodology has also been developed, which consists of the core scaling and loop scaling [1].

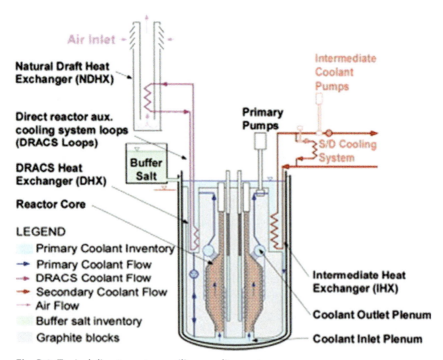

Fig. B.1 Typical direct reactor auxiliary cooling system.

B.2 Decay heat removal system in various reactor designs

As Clinch River Breeder Reactor Project designed by Westinghouse around 1970s time frame on their liquid metal fast breeder reactor (see Fig. B.2), there are three backup systems to remove decay heat in case of nonavailability of the normal heat sink.

1. The first one is the protected air-cooled condenser system that cools the steam drum directly.
2. A second heat sink can be made available by opening the safety relief valve in steam line, thereby venting steam to atmosphere.

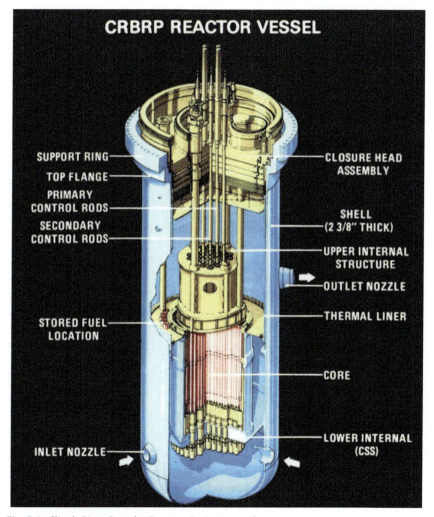

Fig. B.2 Clinch River Breeder Reactor Project vessel.

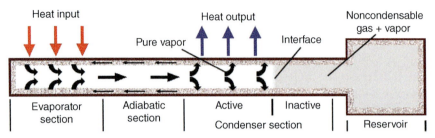

Fig. B.3 Variable conductance heat pipe infrastructure depiction.

3. The third system is a completely separate overflow heat removal system provided to extract heat directly from the in-vessel primary loop. The heat sink for this system is provided by an air-cooled heat exchanger. However, after episode of There Mile Island around 1978 the first author of this book (Zohuri), while was working at Westinghouse Advanced Reactor in Waltz Mill, suggested installation of Mercury heat pipe as an alternative for the heat exchanger and designed series of these heap pipes if form of variable conductance one as an alternative (see Fig. B.3) [2].

The third system in above provides a decay heat removal (DHR) in situations where the steam generator is not available. The DHR system of a loop-type such as Japan Sodium-cooled Fast Reactor consists of a combination of one loop of a DRACS and two loops of a primary reactor cooling system (PRACS) as shown in Fig. B.4.

The heat exchanger (HX) of DRACS is located in the reactor vessel's upper plenum. Each heat exchanger of PRACS is located in intermediate heat exchanger (IHX) upper plenum. These systems operate fully by natural convection and are activated by opening of dc-power-operated dampers and the cold pool are thermally coupled by the PRACS, which is composed of heat exchangers, fluidic diodes, and connecting pipes as shown in Fig. B.5. The fluidic diode reduces leakage flows under primary loop forced circulation.

As part of Generation IV (GEN-IV), the two of the main features of the advanced liquid metal reactor (ALMR) designs are a large heat capacity sodium pool and the utilization of DHR system that operate on natural convection. The use of highly reliable passive DHR system such as DRACS, combined with heat pipe (HP), either fixed or variable type can significantly reduce the risk profile of an ALMR as a fully inherent shutdown system, when we do not require an actor in the loop to take steps of reactor shutdown due to a manmade of natural disaster and accident.

Direct reactor auxiliary cooling system (DRACS) 257

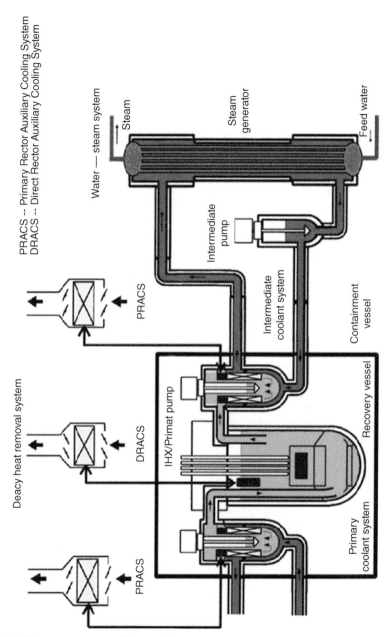

Fig. B.4 Primary reactor auxiliary cooling system.

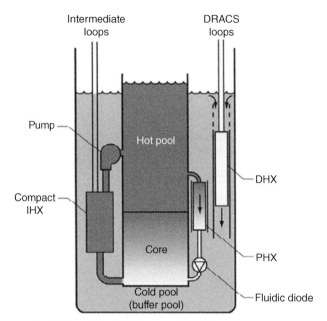

Fig. B.5 Innovative design of the DHR system depiction.

To accurately predict the behavior of the reactor sodium pool during the operation of these types of reactor systems—and consequently their performance—three-dimensional thermal-hydraulic commercial codes such as COMMIX are needed [3].

More specifically, such codes are necessary to predict the flow through the core, the potential for thermal stratification in the hot and cold pols, and the effectiveness of the large thermal mass of the sodium pool in mitigating the decay power driven thermal transient during the early time period (≈ 20 h), when decay heat exceeds on the heat loss. In addition, a detailed thermal analysis is required to evaluate the stresses on the structural components of the system [3].

The objective of such approach is to analyze a DRACS test to support the validation of any computational code either in house written one or a commercial one such as COMMIX and the design of ALMRs. This is one of a series tests that one can perform to generate the thermal-hydraulic data for the validation of these type of codes and to demonstrate the performance of the reactor vessel auxiliary cooling system (RVACS) and DRACS of ALMR designs like sodium advanced fast reactor (SAFR) and power reactor inherently safe module (PRISM) as part of safety and commercial licensing of such production power plants (see Fig. B.6).

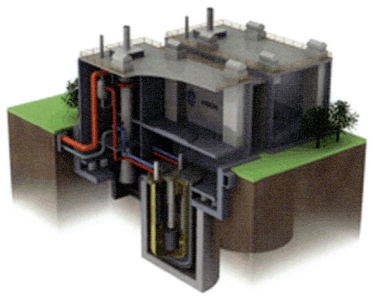

Fig. B.6 Drawing of the PRISM reactor.

Note that: PRISM (sometimes S-PRISM from Super PRISM) is the name of a nuclear power plant design by GE Hitachi Nuclear Energy (GEH).

B.3 Experimental validation of passive decay heat removal technology for FHR

Like other reactors, FHRs require decay heat cooling systems. A DRACS can be used to transfer heat from the reactor coolant to the atmosphere. In a sodium-cooled reactor it is typically a sodium loop with primary sodium providing the decay heat through a heat exchanger to the natural circulation DRACS sodium loop that dumps its heat to a sodium-to-air heat exchanger (Fig. B.7).

In an FHR there are added challenges. First, the reactor peak temperature is typically near 700 °C. Second, the mainline FHR coolant is a molten salt with a freezing point of about 460 °C. It is important not to freeze the primary coolant because that could stop circulation through the reactor core that then could result in high temperatures in the reactor core with fuel failure. Third, neutron interaction with the salt coolant generates tritium that can diffuse through metal heat exchangers—escaping to the

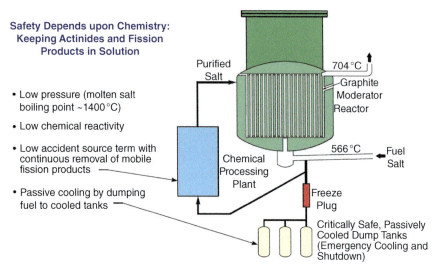

Fig. B.7 MSRs decay heat approach.

environment via DRACS. Last, whatever the fluid in the DRACS, one must consider chemical compatibility with the molten salt in the even there is a heat exchanger failure.

One solution that we as authors of this book, can propose is utilization of multiple heat pipes either fixed or variable type depending on heat duty transport that these pipes need to manage.

The main-line heat-pipe coolant options for DRACS would be sodium and potassium-but a survey of fluoride salts would also be conducted (initial analysis has not found likely candidates, but a more in-depth search would be made). A vertical heat pipe has a fluid that boils, the vapors travel upward, the vapors are condensed and the liquid flows back down the pipe walls. Heat pipes have the characteristic that they can be designed to increase heat transfer by more than an order of magnitude over a small temperature range-what we want for a FHR DRACS, where there is very low heat losses at low temperatures and high heat removal rates above a preset temperature. NASA has developed sodium, potassium and cesium heat pipes for proposed space reactors with one of the goals to "turn-on" above a preset temperature. Such a system would require many parallel heat pipes to remove the desired quantities of heat providing redundancy. That is a major advantage for this application. It implies that the inventory of coolant in any heat pipe is small and thus potential quantities of sodium or potassium that could go into the primary system if heat pipe failure would be small. The addition of sodium or potassium metal into the primary coolant would change the chemical redox (primary concern) as well as some neutronic impacts.

Separate from the DRACS heat transfer is the requirement to prevent tritium diffusion through metal pipes to the environment. The base-case tritium barrier will be a double-wall heat exchanger purged with an inert gas containing small quantities of oxygen. The oxygen would convert any tritium to $3H_2O$ that does not diffuse through metal walls. An oxide tritium barrier may be included where the low oxygen level helps preserve the tritium barrier. The backup is a tritium getter between the walls-an option that has been investigated for some systems. Tradeoff studies will be done before selection of what combination of methods will be used for tritium control.

It is worth to mention that; nuclear reactor power systems could revolutionize space exploration and support human outpost on the Moon and Mars. This paper reviews current static and dynamic energy conversion technologies for use in space reactor power systems and provides estimates of the system's net efficiency and specific power, and the specific area of the radiator. The suitable combinations of the energy conversion technologies and the nuclear reactors classified based on the coolant type and cooling method, for best system performance and highest specific power, are also discussed. In addition, four space reactor power system concepts with both static and dynamic energy conversion are presented. These systems concepts are for nominal electrical powers up to 110 kW_e and have no single point failures in reactor cooling, energy conversion and heat rejection. Two power systems employ liquid metal heat pipes cooled reactors, thermoelectric (TE) and alkali-metal thermal-to-electric conversion (AMTEC) units for converting the reactor power to electricity, and potassium heat pipes radiators. The third power system employs SiGe TE converters and a liquid metal cooled reactor, with a core divided into six identical sectors. Each sector has a separate energy conversion loop, a heat rejection loop, and a rubidium heat pipes radiator panel. The fourth power system has a gas cooled reactor, with a sectored core. Each of the three sectors in the core is coupled to a separate closed Brayton cycle (CBC) loop with He–Xe (40 g/mol) working fluid and a NaK-78 secondary loop, and two separate water heat pipes radiator panels [4].

In summary, The DRACS is a passive residual heat removal system proposed for the FHR that combines the coated particle fuel and graphite moderator with a liquid fluoride salt as the coolant. The DRACS features three natural circulation/convection loops that rely on buoyancy as the driving force and are coupled via two heat exchangers, namely, the DHX and the NDHX.

In case of MSRs, these reactors have a different DH approach as illustrated in Fig. B.4, by dumping fuel salt to tanks. And most probably it utilizes the HP as passive heat transfer system for its decay heat issues.

References

[1] B. Zohuri, Dimensional Analysis and Self-Similarity Methods for Engineers and Scientists, first ed., Springer, New York, 2015.
[2] B. Zohuri, Heat Pipe Design and Technology: Modern Applications for Practical Thermal Management, second ed., Springer, New York, NY, 2016.
[3] B. Zohuri, Thermal-Hydraulic Analysis of Nuclear Reactors, second ed., Springer, New York, 2017.
[4] M El-Genk, Space nuclear reactor power system concepts with static and dynamic energy conversion", Energy Convers. Manage. 49 (3) (2008) 402–411.

APPENDIX C

Heat pipe general knowledge

A heat pipe is a two-phase heat transfer device with a very high effective thermal conductivity. It is a vacuum tight device consisting of an envelope, a working fluid, and a wick structure. As shown in Fig. C.1, the heat input vaporizes the liquid working fluid inside the wick in the evaporator section. The saturated vapor, carrying the latent heat of vaporization, flows toward the colder condenser section. In the condenser, the vapor condenses and gives up its latent heat. The condensed liquid returns to the evaporator through the wick structure by capillary action. The phase change processes and two-phase flow circulation continue as long as the temperature gradient between the evaporator and condenser are maintained.

Heat pipes function by absorbing heat at the evaporator end of the cylinder, boiling and converting the fluid to vapor. The vapor travels to the condenser end, rejects the heat, and condenses to liquid. The condensed liquid flows back to the evaporator, aided by gravity. This phase change cycle continues as long as there is heat (i.e., warm outside air) at the evaporator end of the heat pipe. This process occurs passively and there is no external electrical energy required.

At the hot interface of a heat pipe a liquid in contact with a thermally conductive solid surface turns into a vapor by absorbing heat from that surface. The vapor then travels along the heat pipe to the cold interface and condenses back into a liquid—releasing the latent heat. The liquid then returns to the hot interface through either capillary action, centrifugal force, or gravity, and the cycle repeats. Due to the very high heat transfer coefficients for boiling and condensation, heat pipes are highly effective thermal conductors. The effective thermal conductivity varies with heat pipe length and can approach 100 kW/(m K) for long heat pipes, in comparison with approximately 0.4 kW/(m K) for copper.

Heat pipes employ evaporative cooling to transfer thermal energy from one point to another by the evaporation and condensation of a working fluid or coolant. Heat pipes rely on a temperature difference between the ends of the pipe and cannot lower temperatures at either end below the ambient temperature (hence they tend to equalize the temperature within the pipe; Fig. C.2).

Heat pipes have an envelope, a wick, and a working fluid. Heat pipes are designed for very long-term operation with no maintenance, so the heat pipe

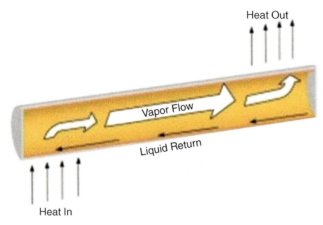

Fig. C.1 A simple physical configuration of heat pipe.

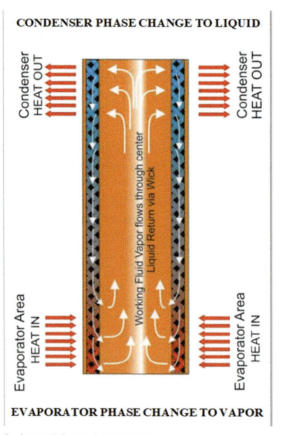

Fig. C.2 Internal schematic heat pipe structure.

wall and wick must be compatible with the working fluid. Some material/working fluids pairs that appear to be compatible are not. For example, water in an aluminum envelope will develop large amounts of noncondensable gas over a few hours or days, preventing normal operation of the heat pipe.

C.1 Heat pipe materials and working fluids

Since heat pipes were rediscovered by George Grover in 1963, extensive life tests have been conducted to determine compatible envelope/fluid pairs, some going on for decades. In a heat pipe life test, heat pipes are operated for long periods of time, and monitored for problems such as noncondensable gas generation, material transport, and corrosion.

C.2 Different types of heat pipes

In addition to standard, constant conductance heat pipes, there are a number of other types of heat pipes, including:
- Vapor chambers (planar heat pipes), which are used for heat flux transformation, and isothermalization of surfaces.
- Variable conductance heat pipes, which use a noncondensable gas to change the heat pipe effective thermal conductivity as power or the heat sink conditions change.
- Pressure controlled heat pipes, which are a variable conductance heat pipe where the volume of the reservoir, or the noncondensable gas mass can be changed, to give more precise temperature control.
- Diode heat pipes, which have a high thermal conductivity in the forward direction, and a low thermal conductivity in the reverse direction.
- Thermosyphons, which are heat pipes where the liquid is returned to the evaporator by gravitational/accelerational forces.
- Rotating heat pipes, where the liquid is returned to the evaporator by centrifugal forces.

C.3 Nuclear power conversion

Grover a scientist at Los Alamos National Laboratory and his colleagues were working on cooling systems for nuclear power cells for space craft, where extreme thermal conditions are encountered. These alkali metal heat pipes transferred heat from the heat source to a thermionic or thermoelectric converter to generate electricity.

Since the early 1990s, numerous nuclear reactor power systems have been proposed using heat pipes for transporting heat between the reactor core and the power conversion system. The first nuclear reactor to produce electricity using heat pipes was first operated on September 13, 2012 in a demonstration using flattop fission.

In nuclear power plant application, heat pipes can be used as a passive heat transfer system for performing as overall thermal hydraulic and natural circulation subsystem in an inherent shutdown, heat removal system in the core (i.e., installed on top of the core doom) of nuclear reactor such as molten salt or liquid metal fast breeder reactor type, as a secondary fully inherent shutdown system loop acting like heat exchanger from safety point of view so the reactor never reaches to its melting point in case of accidental events.

C.4 Benefits of heat pipe devices

The benefits of these devices are listed below as:
- High thermal conductivity (10,000–100,000 W/m K)
- Isothermal
- Passive
- Low cost
- Shock/vibration tolerant
- Freeze/thaw tolerant

C.5 Limitations

The limitations of heat pipes are listed as follows:
- Heat pipes must be tuned to particular cooling conditions. The choice of pipe material, size and coolant all have an effect on the optimal temperatures at which heat pipes work.
- When used outside of its design heat range, the heat pipe's thermal conductivity is effectively reduced to the heat conduction properties of its solid metal casing alone—in the case of a copper casing, around 1/80 of the original flux. This is because below the intended temperature range the working fluid will not undergo phase change; and above it, all of the working fluid in the heat pipe vaporizes and the condensation process ceases.
- Most manufacturers cannot make a traditional heat pipe smaller than 3 mm in diameter due to material limitations.

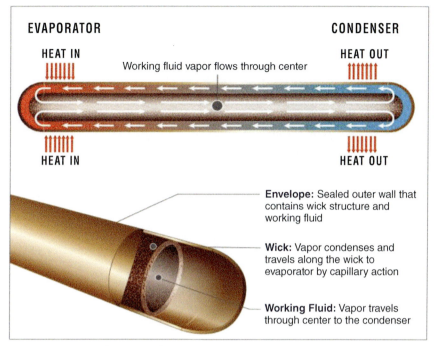

Fig. C.3 Tope view depiction of heat pipe.

C.6 Conclusion

Overall, a heat pipe is a heat-transfer device that combines the principles of both thermal conductivity and phase transition to effectively transfer heat between two solid interfaces (Fig. C.3).

Phase-change processes and the two-phase flow circulation in the HP will continue as long as there is a large enough temperature difference between the evaporator and condenser sections. The fluid stops moving if the overall temperature is uniform but starts back up again as soon as a temperature difference exists. No power source (other than heat) is needed.

In some cases, when the heated section is below the cooled section, gravity is used to return the liquid to the evaporator. However, a wick is required when the evaporator is above the condenser on earth. A wick is also used for liquid return if there is no gravity, such as in NASA's micro-gravity applications.

C.7 Control

Heat pipe heat exchangers can be designed with proportional control to operate in conjunction with the climate control system.

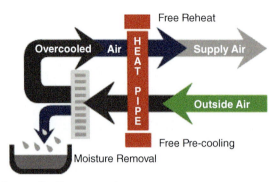

Fig. C.4 Heat pipe application concepts.

C.8 Engineering

Advanced Cooling Technologies Corporation is one of the American Heat Pipe manufacturer that designs and built the heat pipe.

All Advanced Cooling Technologies heat pipe heat exchangers are custom built to accurately meet and exceed customer expectations.

C.9 Heat pipe applications

A heat pipe is a passive energy recovery heat exchanger that has the appearance of a common plate-finned water coil except the tubes are not interconnected. Additionally, it is divided into two sections by a sealed partition. Hot air passes through one side (evaporator) and is cooled while cooler air passes through the other side (condenser). While heat pipes are sensible heat transfer exchangers, if the air conditions are such that condensation forms on the fins there can be some latent heat transfer and improved efficiency (see Fig. C.4).

Heat pipes are tubes that have a capillary wick inside running the length of the tube, are evacuated and then filled with a refrigerant as the working fluid and are permanently sealed. The working fluid is selected to meet the desired temperature conditions and is usually a Class I refrigerant. Fins are similar to conventional coils - corrugated plate, plain plate, spiral design. Tube and fin spacing are selected for appropriate pressure drop at design face velocity. HVAC systems typically use copper heat pipes with aluminum fins; other materials are available.

C.9.1 Heat pipe advantages

The use of the heat pipe advantages are:
- Passive heat exchange with no moving parts,
- Relatively space efficient,

- The cooling or heating equipment size can be reduced in some cases,
- The moisture removal capacity of existing cooling equipment can be improved,
- No cross-contamination between air streams.

C.9.2 Heat pipe disadvantages

The use of the heat pipe disadvantages are:
- Adds to the first cost and to the fan power to overcome its resistance,
- Requires that the two air streams be adjacent to each other,
- Requires that the air streams must be relatively clean and may require filtration.

C.9.3 Applications

Heat pipe heat exchanger enhancement can improve system latent capacity. For example, a 1 °F dry bulb drop in air entering a cooling coil can increase the latent capacity by about 3%. Both cooling and reheating energy is saved by the heat pipe's transfer of heat directly from the entering air to the low-temperature air leaving the cooling coil. It can also be used to precool or preheat incoming outdoor air with exhaust air from the conditioned spaces.

C.9.4 Best applications

- Where lower relative humidity is an advantage for comfort or process reasons, the use of a heat pipe can help. A heat pipe used between the warm air entering the cooling coil and the cool air leaving the coil transfers sensible heat to the cold exiting air, thereby reducing or even eliminating the reheat needs. Also, the heat pipe precools the air before it reaches the cooling coil, increasing the latent capacity and possibly lowering the system cooling energy use.
- Projects that require a large percentage of outdoor air and have the exhaust air duct in close proximity to the intake, can increase system efficiency by transferring heat in the exhaust to either precool or preheat the incoming air.

C.9.5 Possible applications

The possible applications are:
- Use of a dry heat pipe coupled with a heat pump in humid climate areas.
- Heat pipe heat exchanger enhancement used with a single-path or dual-path system in a supermarket application.

- Existing buildings where codes require it or they have "sick building" syndrome and the amount of outdoor air intake must be increased.
- New buildings where the required amount of ventilation air causes excess loads or where the desired equipment does not have sufficient latent capacity.

C.9.6 Technology types and resources

Hot air is the heat source, flows over the evaporator side, is cooled, and evaporates the working fluid. Cooler air is the heat sink, flows over the condenser side, is heated, and condenses the working fluid. Vapor pressure difference drives the evaporated vapor to the condenser end and the condensed liquid is wicked back to the evaporator by capillary action. Performance is affected by the orientation from horizontal. Operating the heat pipe on a slope with the hot (evaporator) end below horizontal improves the liquid flow back to the evaporator. Heat pipes can be applied in parallel or series.

C.9.7 Efficiency

Heat pipes are typically applied with air face velocities in the 450–550 feet per minute range, with 4–8 rows deep and 14 fins per inch and have an effectiveness of 45%–65%. For example, if entering air at 77 °F is cooled by the heat pipe evaporator to 70 °F and the air off the cooling coil is reheated from 55 °F to 65 °F by the condenser section, the effectiveness is 45% [=(65-55)/(77-55) = 45%]. As the number of rows increases, effectiveness increases but at a declining rate. For example, doubling the rows of a 48% effective heat pipe increases the effectiveness to 65%.

Tilt control can be used to:
- change operation for seasonal changeover,
- modulate capacity to prevent overheating or overcooling of supply air,
- decrease effectiveness to prevent frost formation at low outdoor air temperatures.

Tilt control (six maximum) involves pivoting the exchanger about its base at the center with a temperature-actuated tilt controller at one end. Face and bypass dampers can also be used.

APPENDIX D

Variable electricity and steam-cooled based load reactors

D.1 Introduction

Historically nuclear plants have been designed to produce base-load electricity using steam turbines—the traditional power cycle of the utility industry. Changing markets and changing technologies suggest that air-Brayton power cycles may become the preferred power cycle technology for higher-temperature reactors: fluoride-salt-cooled high-temperature reactors (FHRs), sodium fast reactors and high-temperature gas-cooled reactors (HTGRs). The basis for this conclusion is described herein.

D.2 Implication of low-carbon grid and renewables on electricity markets

In a free market the price of electricity varies with time. Fig. D.1 shows the market price of electricity versus the number of hours per year electricity can be bought at different prices in California (blue bars). Power plants with the lowest operating costs come are dispatched first. As the price of electricity rises, power plants with higher operating costs come online. There are near-zero and negative prices for a significant number of hours per year when electricity production exceeds demand and electricity generators pay the grid to take electricity.

This is a consequence of two effects:
- *Renewables subsidies.* Production tax credits provide revenue for wind and solar plants to produce output independent of electricity demand. An owner of a wind or solar facility will sell electricity into the grid as long as the price paid to the grid to take electricity when there is excess production is less than the subsidy [1].
- *Operational constraints.* Nuclear and fossil plants cannot instantly shut down and restart. They pay the grid at times of negative prices to remain on-line and thus be able to sell electricity a few hours later at high prices.

The addition of significant nondispatchable wind or solar changes the shape of the price curve. The addition of a small amount of solar

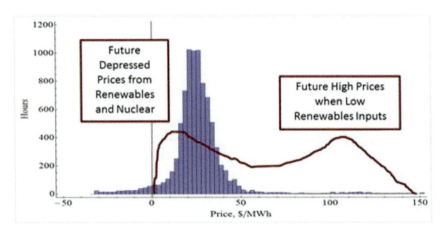

Fig. D.1 Distribution of electrical prices (bar chart), by duration, averaged over CAISO (California) hubs (July 2011–June 2012) and notational price curve (red line) for future low-carbon grid.

is beneficial because the electricity is added at times of peak demand. However, as additional solar is added, it drives down the price of electricity in the middle of sunny days. Each solar owner will sell electricity at whatever price exists above zero. This implies that when 10%–15% of the total electricity demand is met by solar in California, the output from solar systems during midday for parts of the year will exceed electricity demand, the price of electricity will collapse to near or below zero, and the revenue to power plants at these times will collapse to near zero. Each incremental addition of solar at this point lowers the revenue for existing solar electricity producers. The percentage solar is the percentage of all electricity produced by solar—zero in the middle of the night and exceeding electricity demand initially in June in the middle of sunny days. Relatively small fractions of solar have large impacts on prices in the midday but no impact at night when there is no solar.

The same effect occurs as one adds wind capacity, but wind input is more random. As wind penetrates the market it drives the price of electricity down on days with high wind conditions and low electricity demand. Recent studies have estimated this effect in the European market [2, 3]. If wind grows from providing 0% to 30% of all electricity, the average yearly price for wind electricity in the market would drop from 73 €/MWe (first wind farm) to 18€/MWe (30% of all electricity generated). There would be 1000 hours per year when wind could provide the total electricity demand, the price of electricity would be near zero, and 28% of all wind energy would be sold in the market for prices near zero.

The same will occur with nuclear but only when nuclear provides ~70% of the total electricity demand. This is because nuclear plants run at base-load and base-load is about 70% of electricity demand.

In a fossil-fuel dominated system one does not see near-zero prices because fossil plants have low capital costs and high fuel costs. Fossil plants will shut down when electricity prices go below the costs of the fossil fuels. With renewables and nuclear, prices can approach zero for a significant number of hours per year. Without massive subsidies that increase with renewables penetration, this revenue collapse limits solar to ~10% of electricity production and wind to ~20% of electricity production. This also implies that in the long term the price of electricity at times of low renewable input will rise. If other types of power plants operate half the time because they do not generate electricity at times of high renewable inputs, replacement plants will not be built unless there is a rise in the prices of electricity when renewable energy sources are not producing electricity. The red line in Fig. D.1 is a notational price curve one is expected to get if there is large-scale used of renewables with more hours of low-priced electricity (high wind or high solar) and more hours of high-priced electricity with fewer hours of mid-priced electricity. Recent studies on the German grid have reached similar conclusions [4].

The changing shape of the price curve encourages technologies with low capital costs and high operating costs to provide electricity at times of low renewables inputs. The net result is that large-scale wind and solar with existing technologies can result in increased use of fossil fuels to provide electricity at times of low solar or wind conditions. Studies by the State of California [5] and Google [6] have come to similar conclusions.

D.3 Strategies for a zero-carbon electricity grid

There are multiple strategies for a zero-carbon grid; each with specific advantages and disadvantages. Traditionally storage is proposed to meet peak electricity demand; but there are major constraints in the context of a zero-carbon electricity grid. Pumped hydro coupled with nuclear can meet variable electricity demands because the pumped hydro can be charged at night at times of low electricity demand. However, the parallel strategy does not work for systems with large solar or wind inputs. There can be extended periods of low wind or solar that will deplete any storage device and thus storage in systems with large wind or solar components requires backup electricity generation.

There are incentives in coupling storage with nuclear reactors because storing heat is generally cheaper than storing electrons for peak electricity production. Nuclear reactors produce heat and thus can couple efficiently to these systems. Another paper at this conference discusses thermal storage systems coupled to light water reactors [7].

There are seasonal variations in electricity demand. Today the only proposed technology for seasonal storage in a low-carbon grid is hydrogen; but, the round-trip efficiency (electricity to hydrogen to electricity) is less than 50%.

D.4 Nuclear air-Brayton combined cycle strategies for zero-carbon grid

In the last 20 years there have been dramatic improvements in utility gas turbines. Combined cycle efficiency is now ~60%. The cooling water requirements are 40% of a light water reactor because much of the heat is rejected as hot air. For nuclear systems, gas turbines can be run in base-load and peak mode as discussed in the next section. Furthermore, most of the research and development on power cycles worldwide is associated with these power cycles. These dramatic improvements require rethinking what types of power cycles should be coupled to higher-temperature nuclear power reactors.

We describe two sets of studies that couple gas turbines to nuclear reactors. The next section discusses coupling FHRs to gas turbines, followed by a section on coupling sodium-cooled fast reactors to gas turbines. This would not have been a good idea 20 years ago—the technology was not ready.

D.5 Salt-cooled reactors coupled to NACC power system

In the 1950s the United States initialed the Aircraft Nuclear Propulsion Program to develop a jet-powered nuclear bomber. To meet the required temperatures for the jet engine, the United States began development of the molten salt reactor, where the low-pressure liquid fluoride salt coolant was developed to provide high-temperature heat for the jet engines. The development of the Intercontinental Ballistic Missile resulted in the program being cancelled but two test reactors were successfully built.

The program then started development of the molten salt reactor as a power reactor with a steam cycle. The gas turbine technologies of the 1960s could not meet utility requirements. Advances in utility gas turbines over 50 years have now reached the point where it is practical to couple a salt-cooled reactor to a commercial stationary combined-cycle gas turbine.

One advanced salt-cooled reactor system has been proposed to integrate power production with high-temperature heat storage: the FHR with nuclear air-Brayton combined cycle (NACC) and firebrick resistance-heated energy storage (FIRES). The FHR is a new reactor concept [8] that combines (1) a liquid salt coolant, (2) graphite-matrix coated-particle fuel originally developed for high temperature gas-cooled reactors (HTGRs), (3) a NACC power cycle adapted from natural gas combined cycle plants, and (4) FIRES. The FHR concept is a little over a decade old and has been enabled by advances in gas turbine technology and HTGR fuel. The use of solid fuel avoids some of the complications of liquid fuel reactors. The reactor delivers heat to the power cycle between 600 °C and 700 °C. The Chinese plan to build an FHR test reactor by 2020.

A point design for a commercial FHR has been developed with a base-load output of 100 MWe [9]. The power output was chosen to match the capabilities of the GE 7FB gas turbine—the largest rail transportable gas turbine made by General Electric. FHRs with higher output could be built by coupling multiple gas turbines to a single reactor or using larger gas turbines. The development of an FHR will require construction of a test reactor—this size commercial machine would be a logical next step after a test reactor. This point design describes the smallest practical FHR for stationary utility power generation. The market would ultimately determine the preferred reactor size or sizes.

The FHR is coupled to a NACC with FIRES (Fig. D.2). In the power cycle external air is filtered, compressed, heated by hot salt from the FHR while going through a coiled-tube air heat exchanger (CTAH), sent through a turbine producing electricity, reheated in a second CTAH to the same gas temperature, and sent through a second turbine producing added electricity. Warm low-pressure air flow from the gas turbine system exhaust drives a HRSG, which provides steam to either an industrial steam distribution system for process heat sales or a Rankine cycle for additional electricity production. The air from the HRSG is exhausted up the stack to the atmosphere. Added electricity can be produced by injecting fuel (natural gas, hydrogen, etc.) or adding stored heat after nuclear heating by the second CTAH. This boosts temperatures in the compressed gas stream going to the second turbine and to the HRSG.

The incremental natural gas, hydrogen, or stored heat-to-electricity efficiency is 66.4%—far above the best stand-alone natural gas plants

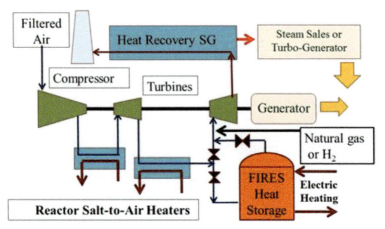

Fig. D.2 Nuclear air-Brayton combined cycle (NACC) with firebrick resistance-heated energy storage (FIRES).

because the added heat is a topping cycle. For comparison, the same GE 7FB combined cycle plant running on natural gas has a rated efficiency of 56.9%. The reason for these high incremental natural gas or stored heat-to-electricity efficiencies is that this high temperature heat is added on top of "low-temperature" 670 °C nuclear heat (Fig. D.3). For a modular 100 MWe FHR coupled to a GE 7FB modified gas turbine that added natural gas or stored heat produces an additional 142 MWe of peak electricity.

The heat storage system consists of high-temperature firebrick heated to high temperatures with electricity at times of low or negative electric prices. The hot firebrick is an alternative to heating with natural gas. The firebrick, insulation systems, and most other storage system components

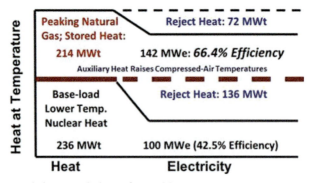

Fig. D.3 Heat and electricity balance for NACC.

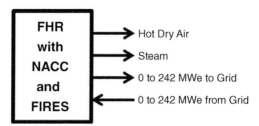

Fig. D.4 Capability of modular FHR with NACC and FIRES.

are similar to high-temperature industrial recuperators. The round-trip storage efficiency from electricity to heat to electricity is ~66%—based on ~100% efficiency in resistance electric conversion of electricity to heat and 66% efficiency in conversion of heat to electricity. That efficiency will be near 70% by 2030 with improving gas turbines.

The plant output is shown in Fig. D.4. When electricity prices are low (less than the price of natural gas), electricity from the FHR is sent to FIRES. In addition, up to 242 MWe of electricity is bought from the grid. The buy capability of the FHR matches the sell capability and thus does not require upgrades to the grid. Because electricity is used to heat the firebrick, firebrick can be heated to 1800°C to minimize the quantity of firebrick required. The hot compressed gas from FIRES is lowered to the turbine limits by either steam injection or mixing with lower-temperature compressed air.

In the existing Texas and California grids, the revenue for an FHR with NACC is 50% higher than a base-load nuclear plant because of the capability to produce more electricity at times of peak demand. This is revenue after subtracting the cost of natural gas for peak electricity production. At natural gas prices 3–4 times the current low prices (natural gas prices in Europe and Asia), the FHR with NACC revenue will be double a base-load nuclear plant.

With FIRES the economics are expected to be dramatically better in the German electricity market today and the expected California market by 2020. In those markets there will be sufficient renewables to drive electricity prices to zero for significant periods of time. It enables replacement of "expensive" natural gas with cheaper electricity. FIRES enables buying massive quantities of electricity when the price is low. Unlike batteries and other electricity storage devices, resistance heaters are inexpensive and thus the system can absorb massive quantities of low-price electricity even if available for short periods of time.

D.6 Sodium-cooled reactors (550 °C) coupled to NACC power system

Since a NACC system looks quite good for a salt-cooled reactor, it is worth considering what it might do for a sodium-cooled reactor. With some modifications it appears that it could be competitive with systems that have been built [10]. A computer model was built based on standard techniques for analyzing Brayton and Rankine systems [11-13]. System performance was optimized by varying the turbine outlet temperatures for a fixed turbine inlet temperature. A second parameter that can usually be varied to obtain optimum performance is the peak pressure in the steam cycle.

For most of the cases considered here this was held constant at 12.4 MPa (1800 psi). Fairly detailed design was attempted for the heat exchangers involved in the system as they tend to dominate system size. The techniques and data were extracted from the text by Kays and London [14].

Consider a baseline system composed of a compressor and two turbines in the standard topping cycle and a steam bottoming cycle with three turbines and two reheat cycles. For a reactor outlet temperature of 550 °C (823 K) and a turbine inlet temperature of 510 °C (783 K), the standard NACC system will achieve an efficiency of 26.2%, at a compressor pressure ratio of 3.1, hardly a competitive number. But if a 95% effective recuperator is added to the system taking the hot exhaust from the HRSG and using it to preheat the compressed gas going into the sodium to air heat exchanger, the efficiency goes up 39.9% and the compressor pressure ratio drops to 1.8.

Since the high-pressure water in the bottoming cycle must be heated and the heating of the air in the air compressor increases the work required, it is possible to split the compressor and add an intercooler that heats the high-pressure water in the bottoming cycle and cools the output from the first part of the compressor. If this is done, the efficiency goes to 40.3 % and the overall compressor pressure ratio goes to 2.0. A system diagram is provided in Fig. D.5.

The ratio of the work in the first half of the compressor to that in the second half is about 1.3:1. For all of these calculations a fairly standard speak steam pressure of 12.4 MPa (1800 psi) was chosen.

The efficiency of the system can be increased by increasing the peak pressure in the steam cycle. This moves the evaporation temperature to a higher level and allows more heat in the HRSG exhaust to preheat the air going into the sodium to air heat exchangers. If the peak steam pressure is increased to 20 MPa (2900 psi) the overall cycle efficiency increases to 41.6% with a compressor pressure ratio of 2.1.

Variable electricity and steam-cooled based load reactors 279

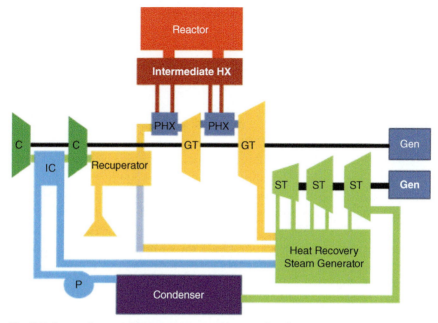

Fig. D.5 System layout with recuperator and intercooler. C, compressor; GT, gas turbine; IC, intercooler; P, pump; PHX, primary heat exchanger; ST, steam turbine.

A final option that was considered was to include two reheat cycles and three turbines in gas turbine. With a recuperator and an intercooler this gave an efficiency of 40.2%. So it is probably not worth this added complexity.

Table D.1 provides a summary of these calculations. The turbine exit temperature was varied to optimize the efficiency of the cycle in all cases. The least accurate number of course is the system volume calculated to estimate relative sizes. But since system size is dominated by heat exchangers, the relative numbers are probably fairly accurate.

Table D.1 Summary of performance calculations.

	TET	CPR	Eff	MdA	WAR	VOL
Baseline	690	3.1	26.2	771	0.036	265
Baseline w/R	740	1.8	39.9	806	0.055	433
Baseline w/R&I	730	2.0	40.3	798	0.051	425
Baseline w/R&I @20 MPa	745	2.1	41.6	801	0.047	445
3 Turbines w/R&I	755	1.8	40.2	751	0.060	438

CPR, compressor pressure ratio; MdA, mass flow rate of air (kg/s); TET, turbine exit temperature (K); VOL, system volume in cubic meters; WAR, water to air mass flow ratio.

The lower temperature sodium reactor can also be augmented by a combustion chamber that burns the air after it has been heated in a nuclear heat exchanger.

At this point the boundaries on what can be done are not well defined. Because the peak temperatures in the nuclear cycle are significantly lower than in current combustion systems, the nuclear systems do not stress turbine blade technologies. Therefore, the nuclear heated systems do not require the use of turbine blade cooling. Thus, it is reasonable to assume that the combustion augmentation will be used with uncooled turbine blades. This sets an upper limit on turbine inlet temperatures of about 1300 K.

That limit is essentially what had limited the power augmentation for a 100 MWe base-load system to 142 MWe of added peak electricity in the case of a salt-cooled reactor. Choosing this same augmentation (142 MW) for a sodium-cooled reactor will require the turbine inlet temperatures to reach 1010 K and the efficiency of burning the added gas is approximately 57.5%.

This is an efficiency that is comparable to what gas combustion systems currently achieve. However, note that there is almost 300 K of additional temperature increase available if desired. Thus, the combustion augmentation of sodium systems could be greater than for higher temperature system. To take this analysis one step further requires more detailed design calculations which are being pursued at this time.

Another interesting aspect of the NACC systems is that they require significantly less circulating water to get rid of waste heat. Consider a 40% efficient system producing 100 MW(e). A closed loop system will have to dump 150 MW(t) to the circulating water system.

The NACC system with recuperator and intercooler will only have to dump 64% of that amount or 96 MW(t). The steam cycle for this type of NACC produces about 64% of the power at an efficiency of 40%.

The air cycle produces 36% of the power. This ratio shifts in favor of the air-Brayton cycle as turbine inlet temperatures increase. At 700 °C the two systems produce about equal amounts of power. The decreased requirement for circulating water should make a larger fraction of the surface of the earth available for power systems installations.

It is also possible to eliminate the steam bottoming cycle entirely at the cost of about 2% in efficiency. This would completely eliminate any dependence on cooling water and allow reactor power system to be placed almost anywhere.

D.7 Power cycle comparisons

The efficiency of NACC power systems continues to increase with increased turbine inlet temperatures. For the foreseeable future there does not appear to be a limitation to using off the shelf materials as it is not likely that a reactor heated system will exceed 1300 K turbine inlet temperature. A comparison of the cycle efficiencies for several cycles that have been proposed for the next generation nuclear plant [15-17] is presented in Fig. D.6.

The calculations for the NACC systems are based on the system described in Fig. D.5 with a peak steam pressure of 12.4 MPa. The data for the other systems were extracted from the referenced publications.

Around 500 °C the NACC cycle falls short of the super critical steam and the super critical CO_2 cycles by 4.1% and 2.6%, respectively. Thus, it will not be competitive on a pure efficiency basis. But by connecting to the large base of components available for gas turbines and reducing the circulating water requirement, it still offers significant advantages.

At the 700 °C point the NACC system has an efficiency that is about 1.4% less than a super critical CO_2 system. However, if the steam bottoming cycle can be operated at 20 MPa rather than 12.4 MPa the efficiency of

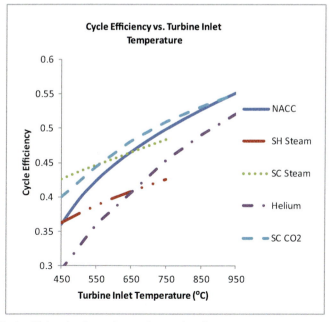

Fig. D.6 Cycle efficiencies for various advanced cycles.

the NACC system surpasses that of the super critical CO_2 system by about 0.5%. So, it is clear that NACC systems can be competitive for most of the next generation nuclear plants that have been proposed.

NACC systems can also be applied to HTGR systems. The major difficulty will be minimizing the size of the helium-to-air heat exchangers. In addition, there is a matching problem with the reheat cycles in the gas turbines. Since the core of an HTGR is very large to provide a good heat transfer area and the temperature increase across the core is usually several hundred degrees, the helium temperature exiting the helium-to-air heat exchangers will be too high to reenter the reactor core directly. So, some efficient technique will be required to use this extra heat before the helium is returned to the reactor core.

This is not an insurmountable problem but will require a slightly different NACC topology for a HTGR like system.

D.8 Conclusions

The electricity market is changing and will continue to change with deregulation, emphasis on a low-carbon grid, and addition of renewables. The change in the market requires rethinking how to best use nuclear energy to meet the needs of the electricity grid—and the need for variable power.

There have been major advances in gas turbines and most of the world's R&D on advanced power cycles is to improve gas turbine technology. Those advances now make it feasible to couple NACC to power reactors. The characteristics of NACC allow base-load reactor operation with variable electricity output—improving economics and enabling a zero-carbon electricity grid. There are significant challenges but large incentives for such power cycles.

NACC systems can be applied to most of the proposed next generation systems. Their strongest competitor in terms of cycle efficiency is the super critical CO_2 system. NACC systems will match or better the efficiency of these systems at or above 700 °C. But NACC systems have the competitive advantage of a large customer base for system hardware, significantly reduced circulating water requirement for rejecting waste heat, and much greater efforts to improve the technology relative to other power cycles.

References

[1] F. Huntowisk, A. Patterson, M. Schnitzer, Negative Electricity Prices and the Production Tax Credit: Why Wind Producers Can Pay Us to Take Their Power—And Why That Is a Bad Thing, The Northbridge Group, New York, NY, 2012.
[2] L. Hirth, The market value of variable renewables, the effect of solar wind power variability on their relative prices, Energy Econ. **38** (2013) 218–236.
[3] L. Hirth, The optimal share of variable renewables: how the variability of wind and solar power affects their welfare-optimal development, Energy J. **36** (1) (2015).
[4] H. Poser, et al., Development and Integration of Renewable Energy: Lessons Learned From Germany, Finadvice, FAA Financial Advisory AG, Switzerland, 2014.
[5] California Council on Science and Technology, California Energy Futures – The View to 2050: Summary Report, 2011.
[6] R. Konningstein, D. Fork, What It Would Really Take to Reverse Climate Change, IEEE Spectrum, 2014, http://spectrum.ieee.org/energy/renewables/what-it-would-really-take-to-reverse-climate-change, October 15, 2020.
[7] C. Forsberg and E. Schneider, "Variable Electricity from Base-load Nuclear Power Plants Using Stored Heat,' Paper 15125, ICAPP, 2015.
[8] C. Forsberg, et al., Fluoride-Salt-Cooled High-Temperature Reactor (FHR) Commercial Basis and Commercialization Strategy. MIT-ANP-TR-153, Massachusetts Institute of Technology, Cambridge, MA, 2014.
[9] C. Andreades, et al., Technical Description of the "Mark 1" Pebble-Bed Fluoride-Salt-Cooled High-Temperature Reactor (PB-FHR) Power Plant, UCBTH-14-002, University of California at Berkeley, Berkeley, CA, 2014.
[10] A.E. Waltars, A.B. Reynolds, Fast Breeder Reactors, Pergamon Press, New York, 1981.
[11] D.G. Wilson, T. Korakianitis, The Design of High-Efficiency Turbomachinery and Gas Turbines, second ed., Prentice Hall, Upper Saddle River, NJ, 1998.
[12] P.P. Walsh, P. Fletcher, Gas Turbine Performance, Blackwell Science, ASME, Fairfield, NJ, 1998.
[13] M.M. El-Wakil, Powerplant Technology, McGraw-Hill, New York, 1984.
[14] W.M. Kays, A.L. London, Compact Heat Exchangers, McGraw Hill, New York, 1964.
[15] U. Oka, S. Koshizuka, Design concept of once-through cycle supercritical-pressure light water cooled reactors, SCR-2000, Proceedings of the First International Symposium on Supercritical Reactors, Tokyo, 2000.
[16] V. Dostal, P. Hejzlar, M.J. Driscoll, The supercritical carbon dioxide power cycle: comparison with other advanced cycles, Nucl. Technol. **154** (2006) 283–301.
[17] B. Zohuri, Innovative Combined Brayton Cycle Systems for the Next Generation Nuclear Power Plants, first ed., Springer, New York, NY, 2014.

APPENDIX E

Variable electricity and steam-cooled based load reactors

E.1 Introduction

In a previous publication, we addressed the possibility of a combined cycle power conversion system for the next generation nuclear plant [3]. This was based on the prediction that several technologies being investigated could provide coolants to a heat exchanger in the region of 1000 K. The higher temperatures offer the opportunity to significantly improve the thermodynamic efficiency of the energy conversion cycle. We jumped to the combined cycle concept because it appears to be in vogue today for natural gas driven power plants. Our results appeared significant in that thermodynamic efficiencies in the range of 45%–50% appeared possible. It also appeared possible to reduce the requirements for the heat dump to the circulating water system significantly, possibly as much as ~75%.

Though these results seemed impressive, they raised some interesting questions. These were:

1. "Since higher temperatures were being considered, couldn't the conventional steam cycle perform as well as the combined cycle?"
2. "Could the circulating water requirement be reduced even further by going to a pure recuperated open air Brayton cycle?"
3. "How would the efficiency of a recuperated Brayton cycle compare with the Combined Cycle efficiencies?"

In order to answer these questions in a realistic manner, detailed system models were developed to include first order designs for all of the appropriate heat exchangers and turbomachinery. As before, this study has concentrated on modeling all power conversion equipment from the fluid exiting the reactor to the energy releases to the environment.

E.2 The recuperated Brayton cycle

For the open air-Brayton cycle, we took a cue from the current practice in steam plants and considered multiple turbines with reheat cycles in between. In our combined cycle analysis, the number of turbines and reheat heat exchangers optimized at the four units level. Using the same models and analysis techniques, a recuperated system appears to optimize out at the two unit level, and it is also reported in Chapter 3 Section 13.3 as well [1].

Obviously, the recuperator can be a very efficient preheater before the first heat exchanger [2-4]. But the pressure drop through the recuperator to get back to atmospheric pressure must be accounted for and the recuperator cannot be designed to minimize only one pressure drop. In the combined cycle analysis, the goal was to minimize the pressure drops in the air systems and tolerate pressure drops in the liquid systems. The liquid pressure drops had very little effect on system efficiency. But a recuperator is a gas to gas heat exchanger [2] and the combined pressure drop is the determining factor.

As a result, the recuperated system with two heater-turbine units was slightly less efficient than the four heater-turbine unit, combined cycle systems (~2%–4%). Since the recuperated Brayton system was simpler with fewer variable parameters, it was chosen to compare with a higher temperature Rankine system [5].

E.3 Modeling the Rankine cycle

The Rankine cycle was modeled in a fairly standard manner with three turbines, two reheat cycles and a condenser and a pump. Feed water heaters were neglected for the sake of simplicity. They would of course add a few per cent efficiency to the simple model used here, but it would be a fairly constant effect that should not vary much with increasing peak temperatures. Each of the reheat processes took the steam back to the peak turbine temperature for the system under consideration. Several system pressures were considered, with the pressure stepping down approximately a factor of four across each of the turbines.

No attempt was made to address any of the material considerations that might be brought on by these higher temperatures in a steam system. Of course, all of the waste heat withdrawn from the condenser would have to be transferred to the circulating water system and returned to the environment. The open cycle Brayton system also has to perform this function by exhausting the working fluid to the environment, but no water supply is required.

E.4 Computer code running results

The most important result is presented in Fig. E.1 where the thermodynamic efficiencies of the recuperated open air Brayton cycle is compared with the efficiencies of potential high temperature Rankine steam cycles.

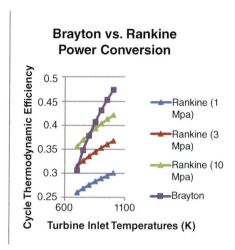

Fig. E.1 Brayton versus power conversion computer run.

The rapid rise in efficiency of the open air Brayton cycle seems to justify its development if the next generation nuclear plant is truly capable of producing the higher temperatures.

The temperature drop across both turbines in the recuperated systems is the same and the turbine exit temperatures are presented in Fig. E.2.

As with all recuperated systems, the compressor ratios for the systems considered here were much lower than in nonrecuperated systems. The optimum pressure ratios all were between 2.0 and 3.0 as depicted in Fig. E.3.

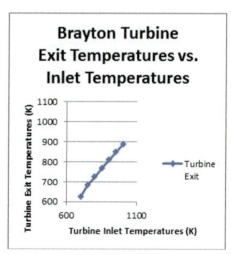

Fig. E.2 Illustration of Brayton turbine exit temperature versus inlet temperature computer model run.

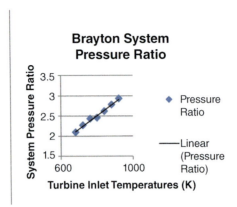

Fig. E.3 Brayton system pressure ratio of computer model run.

E.5 Conclusions

The main conclusion is that open-air Brayton systems appear to be worth investigating, if the higher temperatures predicted for the next generation nuclear plant do materialize. Also recuperated Brayton systems that require no circulating water systems can be installed in nearly every environment on the face of the earth, including deserts.

Note that: The computer model code was developed by this author and Dr. Patrick McDaniel as a joint efforts, while both were at University of New Mexico, Nuclear Engineering Department and it was written in FORTRAN language under steady-state condition in Eulearian frame of reference and increase or decrease in temperature was done increment of 10 degree. The authors of the code still developing transient code model under possibly Lagrangian schema as frame of reference.

References

[1] P. Mc Daniel, C. De Oliviera, B. Zohuri, J. Cole, A combined cycle power conversion system for the next generation nuclear plant. In: ANS Trans., 2012.
[2] W.M. Kays, A.L. London, Compact Heat Exchangers, third ed., Krieger, Malabar, FL, 1998.
[3] D.F. Williams, Assessment of Candidate Molten Salt Coolants for the NGNP/NHI Heat-Transfer Loop, ORNL/TM-2006/69, Oak Ridge National Laboratory, New York, NY, 2006.
[4] J. Mattingly, H. Von Ohain, Elements of Propulsion: Gas Turbines and Rockets, American Institute of Aeronautics and Astronautics, New York, NY, 2006.
[5] El-Wakil, M.M., Powerplant Technology, McGraw-Hill (198), New York, 2015.

Index

Page numbers followed by "*f*" and "*t*" indicate, figures and tables respectively.

A

Advanced high-temperature reactor (AHTR), 6
Advanced liquid metal reactor (ALMR), 256
Advanced small modular reactor (AdvSMR), 1, 24
Air-Brayton Cycle, 87
Aircraft nuclear propulsion (ANP), 1–2
Aircraft reactor experiment (ARE), 1–2
Alkali-metal thermal-to-electric conversion (AMTEC), 261
Alternating current (AC), 39*t*
Area density, 107
Argonne National Laboratory (ANL), 14–15
Atomic Energy Commission (AEC), 2, 14
Atomic Energy Research Establishment (AERE), 10
Atomic waste burner (AWB), 11–12

B

Beryllium oxide (BeO), 9
Bismuth-graphite reactor (BGR), 2
Brayton cycle, 87, 95
Brayton-Rankine power conversion system, 134*t*

C

Canadian Nuclear Safety Commission (CNSC), 11
Centre national de la recherchescientifique (CNRS), 12
Clinch River Breeder Reactor Project (CRBRP), 255
Closed Brayton Cycle (CBC), 261
Closed circuit, 116
Coiled-tube air heat exchanger (CTAH), 5
Combined cycle (CC), 176
Combined cycle efficiency,
Combined cycle gas turbine (CCGT), 116
Compact molten salt reactor (CMSR), 12
Compressor pressure ratio (CPR), 149, 182*t*
Cooperative Research and Development Agreement (CRADA), 6
Cost of electricity (COE), 31, 86–87, 154

D

Decay heat removal (DHR), 40, 225
Denatured molten salt reactor (DMSR), 9–10, 43
Department of Energy (DOE), 1, 121
Direct reactor auxiliary cooling system (DRACS), 208
Division of Reactor Development (DRD), 2, 259–260
DRACS Heat Exchanger (DHX), 208, 253
Dual cycle, 132–133
Dual Fluid Reactor (DFR), 8, 12

E

Effectiveness method, 98, 98–99
Emergency core cooling system (ECCS), 39*t*
Environmental Program Requirements (EPR), 125
Experimental Breeder Reactor II, 218

F

Firebrick resistance-heated energy storage (FIRES), 206–207, 276*f*
Fluid fuels reactor (FFRs), 2
Fluoride salt- cooled High-temperature Reactors (FHRs), 5, 13, 200

G

Gas turbine combined cycle (GTCC), 142*f*, 173*f*
General electric (GE), 2, 130
Generation Four Initiative (GEN-IV), 13
Generation IV international forum (GIF), 9–10
Generation-III (GEN-III), 46–47
Generation-IV (GEN-IV), 13, 37

289

H

Hastelloy-N, 3
Haynes 230 envelope, 218
Heat exchanger (HX), 150, 153
Heat exchanger counter flow, 14–15, 17
Heat exchanger parallel flow, 96
Heat Pipe (HP), 198
Heat Recovery Steam Generator (HRSG), 142–143, 172
Henry's law, 221–222
High temperature gas reactor (htGR), 87, 181
High temperature gas-cooled reactors (HTGR), 229, 271
High-level waste (HLW), 27
High-temperature gas-cooled reactors (HTGRs), 31

I

Inconel 600 alloy, 9
Independent power producers (IPPs), 127
Integral molten salt reactor (IMSR), 11, 59
  descriptions, 61
Intercooled combined cycle (IC-CC), 250
Intermediate heat exchanger (IHX), 3, 89
International Atomic Energy Agency (IAEA), 33
International forum (GIF), 9–10

J

Japan Sodium-cooled Fast Reactor (JSFR), 256

L

Light water reactor (LWR), 69
Liquefied petroleum gas (LPG), 85
Liquid fluoride reactors (LFR), 8
Liquid fluoride thorium reactor (LFTR), 9, 10–11
Liquid metal fast breeder reactor (LMFBR), 4, 247
Liquid-salt very-high-temperature reactor (FHR), 8, 13
Logarithmic mean temperature difference (LMTD), 98
Low-enriched uranium (LEU), 1
Low-level wastes (LLWs), 234–235

M

Merit number, 212
Mixed/unmixed flow, 97f
Mixed oxide (MOX), 26
Molten chloride fast reactors (MCFRs), 204, 230
Molten salt breeder experiment (MSRE), 3
Molten salt breeder reactor (MSBR), 4–5, 9
Molten Salt Converter Reactor (MSCR), 3
Molten salt fast reactor (MSFR), 10
Molten salt reactor (MSR), 1, 2, 60f
Molten salt reactor experiment (MSRE), 8, 9, 78f
Molten salt breeder reactor (MSBR), 4–5, 9
Molten-Salt Reactor Experiment (MSRE), 8, 9

N

Net positive suction head (NPSH), 39t, 59
Next generation nuclear plant (NGNP), 86–87
Non-nuclear weapons states (NNWS), 33
Non-Proliferation Treaty (NPT), 33
Nuclear air combined cycle (NACC), 52
Nuclear Air-Brayton Combined Cycle (NACC), 7f, 171
Nuclear Air-Brayton Recuperated Cycle (NARC), 140
Nuclear Energy for the Propulsion of Aircraft (NEPA), 1, 145f
Nuclear Non-Proliferation (NNP), 34
Nuclear Weapon States (NWS), 33
Number of transfer units (NTU), 98
Number of transfer units (NTU), 98

O

Oak Ridge National Laboratories (ORNL), 1, 59, 65
Office of Nuclear Energy, 121
Office of Science and Technology, 5
Open circuit, 116
Overflow Heat Removal System (OHRS), 256

P

Pebble Bed Fluoride-salt-cooled High-temperature Reactor (PB-FHR), 5

Power Conversion System (PCS), 140, 141
Power Reactor Inherently Safe Module (PRISM), 256
Prandtl number, 216
Pressurized water reactor (PWR), 138, 219
Primary reactor cooling system (PRACS), 256
Protected air-cooled condenser (PACC), 255

R
Radar Ocean Reconnaissance Satellite (RORSAT), 26, 27f
Rankine bottoming cycle, 131–132, 132
Rankine cycle, 85, 93, 114, 133
Rankine system, 166t
Reactor Protection System (RPS), 29
Reactor Vessel Auxiliary Cooling System (RVACS), 258
Recuperated Brayton Cycle (RBC), 247
Regenerated Mixture (REMIX), 26–27
Research and Development (R&D), 3–4, 6
Revolution per minute (RPM), 73

S
Safety Assessment of the Molten Salt Fast Reactor (SAMOFAR), 12
Savannah River Site, 242
Shanghai Institute of Applied Physics (SINAP), 6
Small modular reactor (SMR), 11
Sodium advanced fast reactor (SAFR), 258
Sodium-cooled Fast Reactors (SFRs), 274
Solar power (CSP), 211–212

Stable salt reactor (SSR), 8, 13
Steam Generator (SG), 192
Steam assisted gravity drainage (SAGD), 73
Steam-Rankine Cycle, 172
Stirling Space Power Converter, 218
Supercritical (SC) water cycle, 191

T
Terrestrial Engineering (TE) of Canadian Company, 75
Thermal-neutron reactor, 67
Thermal storage and desalination, 79
Thorium molten-salt reactor (TMSR), 12–13
Thorium molten salt reactor-liquid fuel (TMSR-LF), 12–13
Thorium molten salt reactor—solid fuel 1 (TMSR-SF1), 6, 13
Turbine-generator systems, 73

U
U S Department of Energy's Office of Science, 121
US Nuclear Regulatory Commission, 120, 137–138

V
Very High Temperature Reactor (VHTR), 13, 25f

W
Waste Management, 74
Westinghouse, 24

Printed in the United States
by Baker & Taylor Publisher Services